Property of
Burney Scoles

INTRODUCTION TO MICROWAVES

PRENTICE-HALL INTERNATIONAL, INC. *London*
PRENTICE-HALL OF AUSTRALIA, PTY., LTD. *Sydney*
PRENTICE-HALL OF CANADA, LTD. *Toronto*
PRENTICE-HALL OF INDIA PRIVATE LIMITED *New Delhi*
PRENTICE-HALL OF JAPAN, INC. *Tokyo*
PRENTICE-HALL DE MEXICO, S.A. *Mexico City*

INTRODUCTION TO MICROWAVES

GERSHON J. WHEELER

Sylvania Electronic Systems
a division of
Sylvania Electric Products
a subsidiary of
General Telephone and Electronics, Inc.

Under the editorship of Dr. Irving L. Kosow

PRENTICE-HALL, INC., *Englewood Cliffs, N.J.*

Current printing (last digit):
12 11 10 9 8

Library of Congress Catalog Card Number 63-20979
Printed in the United States of America

PREFACE

In this book I have attempted to present the subject of microwaves without relying on theoretical or mathematical discussions. A knowledge of simple algebra is desirable, but not essential, to use the equations; no calculus or higher mathematics is used. The book first presents what microwave signals are and why they are used, and then discusses microwave equipment, laboratory techniques of measurement, and design.

I would like to acknowledge the assistance of A. L. Tripp, who read and criticized the text and prepared the problems at the end of each chapter. Art work was done by James Cutter. Special thanks, also, are due Mrs. Doris Clift for typing and proof-reading the manuscript.

GERSHON J. WHEELER

CONTENTS

SYMBOLS

A area
a wide dimension of waveguide, radius of inner conductor
a_d radius of inner conductor in presence of dielectric
B susceptance, bandwidth
b narrow dimension of waveguide, radius of outer conductor
C capacitance
c velocity of light
d distance, diameter of round guide
E_1 input voltage
E_o output voltage
E_L^+ incident voltage at load
E_L^- reflected voltage at load
E_R reflected voltage
F noise factor
F_{DB} noise figure
f frequency
f_r repetition frequency
G gain, conductance
H magnetic field
I_1 input current
I_o output current
I_R reflected current
L inductance
L_c conversion loss
l distance, length
N_i input noise
N_o output noise
N_r noise ratio of diode
P_1 input power
P_o output power
P_p peak power
P_A average power
P_R reflected power
pf power factor
prf pulse repetition frequency
Q_E external Q
Q_L loaded Q
Q_U unloaded Q
R resistance, range of antenna
R_b barrier resistance of crystal
r VSWR
r_s spreading resistance of crystal
S_i input signal
S_o output signal
S_{DB} sensitivity
s distance, length
v velocity
v_g group velocity
v_p phase velocity
X reactance
Y_o characteristic admittance
Y_s admittance at a distance s from load
Y_L load admittance
Z_o characteristic impedance
Z_s impedance at a distance s from load
Z_L load impedance

α attenuation constant
α_c conductor attenuation
α_d dielectric attenuation
α_T total attenuation
β phase constant
Γ reflection coefficient
γ propagation constant
δ pulse width, skin depth

tan δ loss tangent $\left(\tan \delta = \dfrac{\epsilon''}{\epsilon'}\right)$

ϵ relative dielectric constant ($\epsilon = \epsilon' + j\epsilon''$)
λ wavelength
λ_c cut-off wavelength
λ_d wavelength in dielectric
λ_g guide wavelength
λ_m minimum wavelength
λ_o wavelength in air
λ_r resonant wavelength of cavity
μ relative permeability
ρ resistivity
ω $2\pi f$

1

BASIC CONCEPTS

The word "microwaves" defines itself: it means very short waves. However, exactly what is meant by "short" depends on who is speaking and his frame of reference. Certainly ultraviolet light has a short wavelength compared to infrared, and 400 cycles per second is a higher frequency (and consequently a shorter wavelength) than 60 cycles per second. All of these waves are forms of electromagnetic energy, but none are microwaves. In general, radio frequencies extend from direct current up to the infrared region. The shortest wavelengths or highest frequencies of the radio spectrum are in the microwave region, but its boundaries are not clearly defined. At its high-frequency edge, it overlaps the infrared. At its low-frequency edge, technique rather than frequency is the determining factor.

The fundamental principles underlying low-frequency radio waves and microwaves are the same. At low frequencies the observed phenomena are easily explained in terms of current flowing in a complete circuit. It is not necessary to use the idea of an electric field and a magnetic field, although these fields exist, and the observed phenomena could, in fact, be described just as well in terms of them. However, at microwave frequencies, it is usually difficult to describe the phenomena in terms of a current.

The engineer accustomed to working at low frequencies thinks in terms of "lumped" circuit elements. A radio circuit has capacitors, inductors, and resistors, which are easy to locate. In a microwave circuit, the inductances and capacitances are "distributed" along a transmission line. It is impossible to point to one spot as the location of a specific circuit element. Instead, every point in the circuit is part of a distributed reactance.

Probably the most important difference between microwave and ordinary radio techniques is the size of components relative to a wavelength. For example, in a waveguide used as a microwave transmission line, the width is greater than half a wavelength. At a frequency of 10,000 megacycles per second, where the wavelength is three centimeters, the waveguide is approximately two and a half centimeters (slightly less than an inch) wide. It is conceivable that a waveguide 90 inches wide could be used with similar results at 100 megacycles. If it were so used, it would be a microwave circuit even though it had a lower frequency than that usually referred to as a microwave. Because wavelength is not a determining factor, there is no sharp demarcation between radio waves and microwaves.

There is a definite relationship between the frequency and the wavelength of an electromagnetic wave. Their product is equal to the velocity of electromagnetic energy, or as it is usually called, *the speed of light.* This is expressed as

$$\lambda f = c \tag{1.1}$$

where λ is wavelength, f is frequency, and c is the speed of light (approximately 3×10^{10} centimeters per second). The value of c is very important and should be memorized in different units; thus,

$$\begin{aligned} c &= 3 \times 10^{10} \text{ cm per sec} \\ &= 1.18 \times 10^{10} \text{ in. per sec} \\ &= 984 \text{ feet per } \mu\text{sec} \end{aligned}$$

(A microsecond is a millionth of a second.) This value of c is the speed of light or of electromagnetic waves in a vacuum. In air, except for very accurate measurements, the same values may be used.

However, in a dielectric medium such as Teflon or Polystyrene, or even in water, the waves are *slowed down* by an amount proportional to the square root of the dielectric constant. Teflon has a dielectric constant of 2.1, and, therefore, the velocity of propagation through teflon will be $(3/\sqrt{2.1}) \times 10^{10}$ centimeters per second. Since the product of the frequency and the wavelength must equal the velocity, the wavelength in a dielectric (for a given frequency) must be shortened also by the square root of the dielectric constant. This is expressed as

$$\lambda_d = \frac{\lambda_0}{\sqrt{\epsilon}} \tag{1.2}$$

where λ_d is the wavelength in the dielectric, λ_0 is the wavelength in air, and ϵ is the dielectric constant of the medium. The frequency f_d remains equal to f_0.

The microwave portion of the electromagnetic spectrum has been set arbitrarily between 300 megacycles (λ = 100 centimeters) and 300,000 megacycles per second (λ = 1 millimeter). The latter is the edge of the infrared region. However, as has been pointed out, this is not a rigid definition but depends on the techniques used. It is customary to omit the words "per second" from the frequency unless there is a chance of confusion.

Microwaves are becoming more and more important in communications, radar, astronomy, navigation, and other fields. The reason lies in two advantages microwaves have over lower frequency signals.

The first advantage is the increased bandwidth. In order to transmit intelligence by radio, it is necessary to transmit a band of frequencies. Ordinary conversations may require a band 1000-cycles wide; music requires 10,000, television a few megacycles. A band of frequencies from 2850 to 3150 megacycles is 300 megacycles wide and thus is capable of transmitting as much intelligence as the whole radio spectrum from direct current to 300 megacycles. This microwave band from 2850 to 3150 megacycles is said to have a bandwidth of ten per cent—its width divided by its center frequency. In general, the narrower the bandwidth, the simpler the circuit.

The second advantage of microwaves is their ability to use high gain, directive antennas. By making antennas several wavelengths wide, which can be done easily when wavelengths are of the order of an inch, it is possible to focus microwave beams in a manner similar to focusing light rays with lenses or reflectors. The wider the aperture of the antenna, in terms of wavelength, the narrower the beam and the higher the gain of the antenna. Thus an antenna 100 centimeters wide operating at 3000 megacycles (λ = 10 centimeters) has an aperture of ten (100/10) wavelengths. It will have a greater gain than an antenna 500 centimeters wide at 300 megacycles, because the latter has an aperture of only five wavelengths.

Microwave energy has a heating effect just as any other form of energy. Because of the short wavelengths and consequent compact circuitry, this heating effect has a variety of nonmilitary applications. Microwave ovens for home cooking heat quickly since the food is cooked by the waves on the inside at the same time as on the outside. Microwave diathermy machines produce heat inside the muscle without cooking the outside. Also, microwave drying machines are used in the printing industry.

QUESTIONS AND PROBLEMS

1.1. What determines the frequency limits of the microwave spectrum?

1.2. Briefly explain the two principal advantages of microwave frequencies over lower frequency radio waves.

1.3. What is the velocity of electromagnetic waves in air? Give the answer in four forms.

1.4. A radar antenna is transmitting at a frequency of 10 Gc. What is the wavelength of the transmitted signal?

1.5. You have an air-filled coaxial transmission line:
   a. What is the velocity of an electromagnetic wave on this line?
   b. The line is now filled with Polystyrene ($\epsilon = 2.5$). What is the new velocity of propagation on the line? What is the wavelength if the frequency is 100 Mc?

1.6. The line being used is *RG*/8U. The frequency of transmission is 500 Mc. What is the wavelength on the line? *ans.* 15.5 in.

2

TRANSMISSION LINES

Radio-frequency lines have several functions. Their obvious use is to convey radio-frequency power from one point to another. Thus, in a radar set at microwaves or in a radio transmitter operating in the broadcast band, a radio-frequency line is used as a transmission channel to carry power from the oscillator or power-amplifier to the antenna. Similarly, transmission lines are used to carry received signals from the antenna back to the receiver. The antenna in low-frequency work is usually placed a short electrical distance from the receiver; that is, the length in wavelengths is small. For example, at a frequency in the middle of the broadcast band, the wavelength is about 1000 feet; a transmission line to the top of a 100 foot tower is only one tenth of a wavelength long. At microwave frequencies, however, with wavelengths expressed in ten inches or less, it is usual for a transmission line to an antenna to be many wavelengths long. This puts more stringent requirements on the physical dimensions of the transmission lines and makes analysis more complicated.

Transmission lines also serve as reactive circuit elements or tuned circuits. When a fixed length of lossless line is short-circuited or open-circuited at the end, it will appear to be a pure reactance which can have any capacitive or inductive value, depending on its length expressed in wavelengths. As a special case, if the length is such that the reactance is infinite (that is, an open circuit), the line can be used as an insulator to support another line. Reactive sections of line are also used as matching elements to cancel unwanted reactances in loads.

A third use for transmission lines is as impedance transformers. For example, when the antenna impedance differs from the characteristic impedance of the transmission line feeding it, the mismatch will cause a reflection of some of the transmitted energy. Consequently, less power will reach the antenna to be radiated. By using a section of line of suitable length and characteristic impedance, it is possible to match the feed-line to the antenna so that all the available power is radiated.

2.1. TYPES OF TRANSMISSION LINES

There are many types of transmission lines used for microwaves, but in general they can be characterized in three groups: (1) the open-wire TEM lines, (2) coaxial TEM lines, and (3) waveguides. Figure 2.1

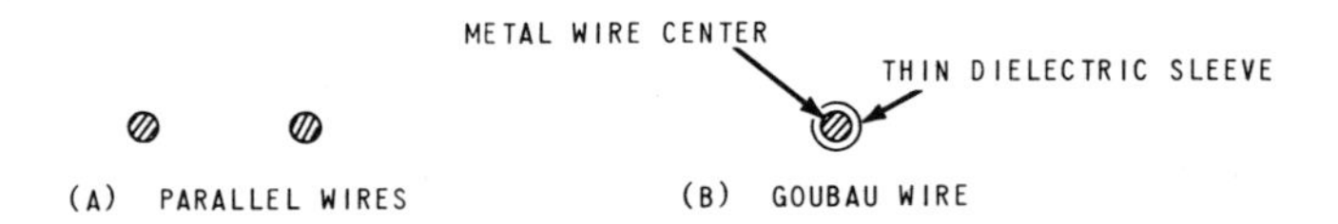

Fig. 2.1. Cross sections of open-wire lines.

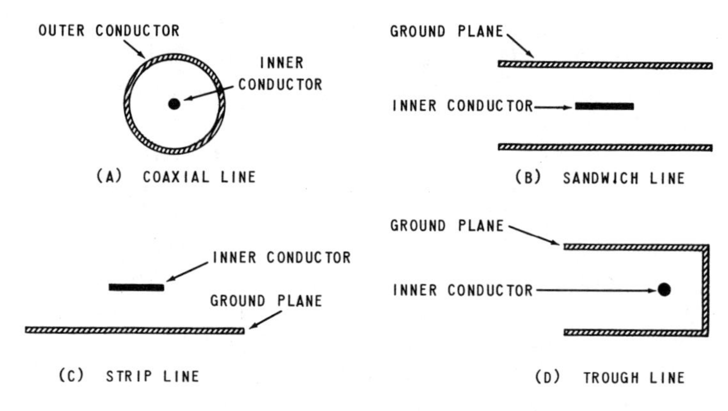

Fig. 2.2. Cross sections of TEM lines.

shows two types of open-wire lines: the common, low-frequency, parallel-wire line and the single-wire Goubau line used at microwave wavelengths. Figure 2.2 shows some types of TEM lines. The coaxial line is the prototype of the TEM, and the other lines are variations or derivatives of it. The inner conductor on all of these may have any cross section. Figure 2.3 shows some types of waveguides. To be sure, all transmission lines guide waves and, therefore, could be called waveguides; however, the term is usually reserved for closed structures which, from a low-frequency point of view, do not look as if they would be able to carry electromagnetic waves. In Fig. 2.3, the first five lines are types of metal guides which are completely enclosed; their cross sections can have almost any shape and still carry energy. The five shown have specific uses at microwaves. The dielectric waveguide and the H-guide (which is a form of dielectric guide with ground plates) have application especially at shorter microwave wavelengths.

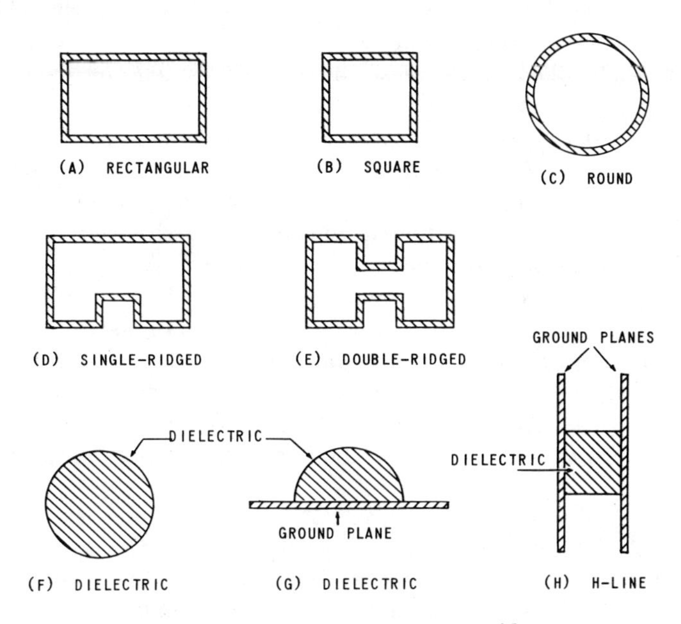

Fig. 2.3. Cross sections of waveguides.

The choice of line is not left to the whim of the designer; it is usually dictated by considerations of electrical, mechanical, and geometrical requirements. Metal waveguides have less loss and can handle greater power than coaxial lines, but they are much larger than coaxial lines at the low microwave frequencies. Open-wire lines are simple but tend to radiate and thus lose some of the power intended for the antenna. Size, weight, machining tolerances, ability to handle power, efficiency, problems in matching, shielding, and reliability all determine the choice of transmission line for a microwave system. Usually, several types are used in one system in order to achieve optimum results.

2.2. LOSSES

There are three kinds of losses in transmission lines: (1) *copper loss,* power dissipated as heat in the metal conductors, (2) *dielectric loss,* also a heat loss in the insulation, and (3) *radiation loss*, power transferred from the transmission line to free space or to near-by circuits. (Power transferred to near-by circuits is sometimes called *induction loss.*) In general, all types of losses are sensitive to frequency changes, increasing as the frequency increases.

Copper loss is expressed in watts and, as in d-c circuits, is equal to I^2R, where I is the rms value of the current in amperes, and R is the resistance of the conductors in ohms. Conductors may be made of any metal but the loss is still called *copper loss* or, sometimes, *I^2R loss*. The resistance, R, however, cannot be measured with an ohmmeter in the usual way, because resistance of metals increases with frequency. This is caused by what is known as *skin effect*. In d-c circuits, the current flows uniformly in the whole cross section of a wire. As the alternating frequency increases, the current tends to flow nearer the surface, or in the "skin," of a conductor. This means that the current is trying to flow in a thinner layer of metal and consequently meets greater resistance. At higher frequencies, such as microwaves, the inner conductor of a coaxial line can be a hollow tube, since only negligible current would flow in the center if it were solid.

Thus the current density is greatest near the surface and it falls off exponentially with depth. The depth at which the current density is $1/e$ (about 37 per cent) of its surface value is called the *skin depth*. Its value in inches is

$$\delta = 2\sqrt{\rho/\mu f} \tag{2.1}$$

where ρ is the resistivity of the metal conductor in ohm-centimeters, μ is its relative permeability, which is unity for most nonmagnetic materials, and f is the frequency in megacycles per second. For copper, $\rho = 1.724 \times 10^{-6}$ ohm-centimeters, and Eq. (2.1) becomes $\delta = 2.61/\sqrt{f}$ mils. At $f = 100$ megacycles, δ is about a quarter of a thousandth of an inch. At higher microwave frequencies δ is so small it is negligible. Thus at microwaves, the thickness of conductors is determined by mechanical considerations, such as rigidity, rather than electrical problems. Table 2.1 presents resistivity values for some common metals used at microwaves.

Table 2.1. RESISTIVITY OF SOME COMMON METALS

Metal	Resistivity
Aluminum	2.827×10^{-6} ohm-cm
Brass	6.724
Copper	1.724
Gold	2.441
Magnesium	4.603
Platinum	10.62
Silver	1.638

Rust or dirt on the inner surface of a waveguide or a coaxial line has little effect on the loss. The high resistance of dirt make its skin depth high so that most of the current still flows in the surface of the metal. Of course, too much dirt can result in dielectric loss.

To keep skin depth low, most microwave lines are made of some easily machinable metal such as brass or a lightweight metal such as aluminum. These lines are then silver plated; the microwave current flows in the highly conductive, low-loss silver layer rather than in the base metal. As long as there is a thin layer of silver, copper, or other highly conductive material on the surface of the base metal, the metal can be replaced by wood, plastic, or anything else, without affecting the microwave circuit.

If a line is air-filled, it has negligible loss in the air. However, other insulating materials do have a dielectric loss which increases with frequency. Commonly used low-loss dielectrics are Teflon, Polystyrene, quartz, and mixtures of these and other materials. Lossy dielectrics are also used in microwaves to absorb power and to prevent it from going where it is not wanted. Radiation loss is not a serious problem at microwaves, since the usual transmission lines are closed waveguides and coaxial lines which are inherently shielded. Radio-frequency transmission lines are generally assumed to be lossless. This is not true, but it is a good enough approximation since lengths are usually kept short and air dielectric or low-loss dielectric is used. If the loss is neglected, circuit analysis is greatly simplified.

2.3. THE INFINITE LINE

A signal fed into a line of infinite length could not reach the far end in a finite time. Consequently, the condition of the far end (e.g., open, short, terminated) can have no effect at the input end. For this reason, transmission line analysis usually begins with an infinite line in order to separate input conditions from output conditions.

If a long line consisting of two parallel, uniform conductors is carrying current, there is a magnetic field around the conductors and a voltage drop along them. The magnetic field, which is proportional to current, indicates that the line has series inductance; the voltage drop indicates the presence of series resistance. Voltage applied across the conductors produces an electrical field between them and charges on them. This indicates that the line contains shunt capacitance and, since the capacitance is never lossless or perfect, some shunt conductance, as well. A unit section of line may then be

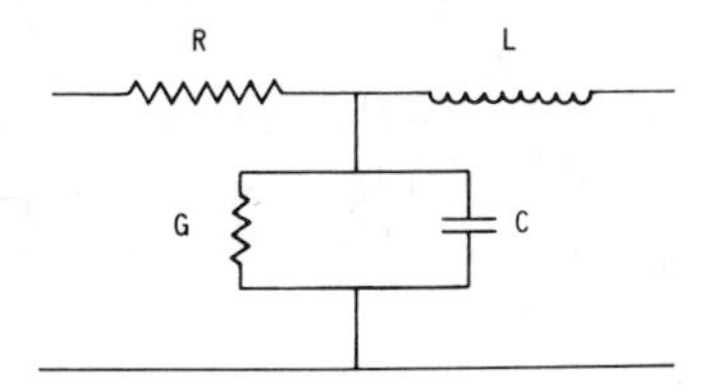

Fig. 2.4. Equivalent circuit of a unit of transmission line.

represented by Fig. 2.4. Actually there are no lumped constants, but instead the resistance, inductance, capacitance, and conductance are distributed along the whole length of the line. The series impedance and shunt admittance of the line are, respectively,

$$\begin{aligned} Z &= R + jX_L \\ Y &= G + jB_c \end{aligned} \tag{2.2}$$

A voltage E_1 applied across the conductors of an infinite line causes a current I_1 to flow. By this observation, the line looks like an impedance which is designated Z_0.

$$Z_0 = \frac{E_1}{I_1} = \sqrt{\frac{Z}{Y}} = \sqrt{\frac{R + jX_L}{G + jB_c}} = \sqrt{\frac{R + j\omega L}{G + j\omega C}} \tag{2.3}$$

The *characteristic impedance* of the line, Z_0, is also referred to as the *surge impedance.*

The propagation of a wave along a line is described completely by $\sqrt{ZY}$. This quantity is called the *propagation constant* and is usually designated γ.

$$\gamma = \sqrt{ZY} = \sqrt{(R + jX)(G + jB)} \tag{2.4}$$

Since γ is complex, it may be defined as

$$\gamma = \alpha + j\beta \tag{2.5}$$

where α (the *attenuation constant*) is the real part of Eq. (2.4) and β (the *phase constant*) is the imaginary part.

2.4. ATTENUATION

Since there is always some loss, no matter how negligible, in a transmission line, the signal traveling along the line becomes attenuated as it progresses; that is, each succeeding voltage peak or current peak is smaller than its predecessor. If the voltage at point 1 is E_1 and the voltage at point 2, farther down the line, is E_2, then

$$\frac{E_1}{E_2} = e^{\alpha l} \tag{2.6}$$

where l is the distance between the points and α is the attenuation constant from Eq. (2.5). In order to make the attenuation directly proportional to distance, a logarithmic unit, the *neper*, is used. From Eq. (2.6),

$$\alpha l = \ln \frac{E_1}{E_2} = 2.303 \log_{10} \frac{E_1}{E_2} \tag{2.7}$$

where αl is measured in nepers, defined by Eq. (2.7). The more common unit of attenuation, however, is the *decibel.*

$$\alpha_{\text{db}} l = 20 \log_{10} \frac{E_1}{E_2} \tag{2.8}$$

Thus, one neper is equal to $20/2.303 = 8.686$ decibels. For example, if the input voltage E_1 on a line is 100 volts and the output voltage E_2 is 80 volts, the attenuation, as measured by Eqs. (2.7) and (2.8), is

$$2.303 \log_{10} \frac{E_1}{E_2} = 0.22 \text{ nepers}$$

$$20 \log_{10} \frac{E_1}{E_2} = 1.94 \text{ db}$$

2.5. PHASE

As mentioned previously, practical transmission lines have such small losses, it may be assumed that the attenuation constant α is zero. This means that all series resistances and shunt conductances are also zero. Therefore, Eqs. (2.2) become

$$\begin{aligned} Z &= j\omega L \\ Y &= j\omega C \end{aligned} \tag{2.9}$$

Substituting for Z and Y Eq. (2.4) becomes

$$\gamma = \sqrt{ZY} = \sqrt{j\omega L \cdot j\omega C} = j\omega \sqrt{LC} \tag{2.10}$$

Since $\gamma = \alpha + j\beta$ and $\alpha = 0$,

$$\beta = \omega \sqrt{LC} \tag{2.11}$$

where β is measured in radians per unit length. The wavelength, λ, is the distance in which the phase changes by 2π radians:

$$\lambda = \frac{2\pi}{\beta} \tag{2.12}$$

The phase velocity $v = \lambda f$ and, therefore,

$$v = \frac{2\pi}{\beta} f = \frac{\omega}{\beta} = \frac{\omega}{\omega \sqrt{LC}} = \frac{1}{\sqrt{LC}} \tag{2.13}$$

Thus, the phase constant β is directly proportional to frequency, and the velocity is independent of frequency.

To illustrate Eqs. (2.10) to (2.13), a typical air-filled coaxial line might have a series inductance of 0.051 microhenries per foot and a shunt capacity of 20.4 picofarads per foot. From Eq. (2.13), the phase velocity is

$$\frac{1}{\sqrt{0.051 \times 10^{-6} \times 20.4 \times 10^{-12}}} = 980 \times 10^{6}$$

measured in feet per second. If the frequency is 2000 megacycles, from Eq. (2.11),

$$\beta = 2\pi \times 2 \times 10^{9} \times 1.02 \times 10^{-9} = 4.08\pi \text{ radians per ft}$$

From Eq. (2.12), the wavelength, is

$$\lambda = \frac{2\pi}{4.08\pi} = 0.49 \text{ feet} = 5.88 \text{ in.}$$

From Eq. (2.14), the characteristic impedance of the line is

$$Z_0 = \sqrt{\frac{0.051 \times 10^{-6}}{20.4 \times 10^{-12}}} = 50 \text{ ohms}$$

It should be noted that the velocity is given by Eq. (2.13). The time required for a signal to travel from one end of the line to the other is equal to the length of the line divided by the velocity. From Eq. (2.13), the time is

$$t = l\sqrt{LC} = \sqrt{(lL)(lC)} \tag{2.13A}$$

where lL is the total inductance and lC is the total capacity. The time t is referred to as the *delay time* of the line. Synthetic lines are built using lumped capacitors and inductors in order to produce a prescribed delay. These synthetic lines are called *delay lines;* their delay times are calculated from Eq. (2.13A).

2.6. CHARACTERISTIC IMPEDANCE

An infinite line looks like an impedance defined by Eq. (2.3), which is, in general, a complex impedance. However, if the line has negligible loss, the values of Z and Y from Eq. (2.9) should be used. Then,

$$Z_0 = \sqrt{\frac{Z}{Y}} = \sqrt{\frac{j\omega L}{j\omega C}} = \sqrt{\frac{L}{C}} \tag{2.14}$$

which is independent of frequency and is real.

If a finite length of line is joined with a similar kind of infinite line, their total input impedance is the same as that of the infinite line itself; for together they make one infinite line. However, the infinite line alone presents an impedance Z_0. It must be concluded that a finite line of characteristic impedance Z_0 has an input impedance Z_0 when it is terminated in Z_0. This is shown in Fig. 2.5.

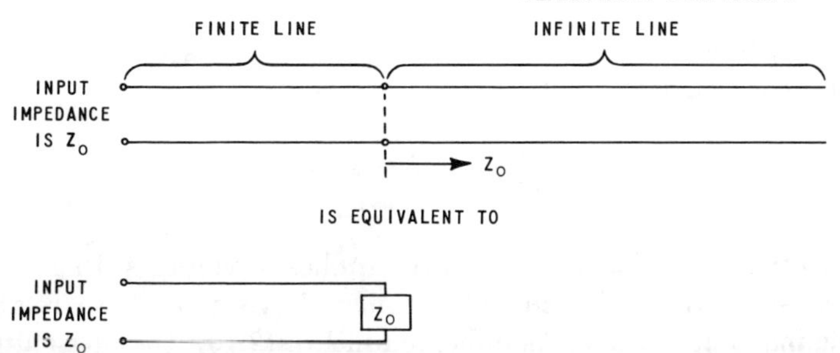

Fig. 2.5. Impedance of matched line.

2.7. REFLECTION COEFFICIENT

When a signal is sent down a transmission line, it travels smoothly until it reaches a discontinuity. Then some of the energy is reflected, the size of the reflection depending on the size and nature of the discontinuity. In this respect, electromagnetic waves behave just as any other waves (e.g., sound or light waves). If there is no discontinuity in the line, the incident wave travels to the receiving end where the termination or load appears as an impedance Z_L. The voltage and current at the receiving, or load, end must be related as follows:

$$E_L = I_L Z_L \tag{2.15}$$

If Z_L is different from Z_0, E_L and I_L cannot be the same as the incident voltage and current which had the relationship $E_1 = Z_0 I_1$. Evidently Z_L looks like a discontinuity and some reflection must result. The voltage E_L across the load Z_L is the sum of an incident voltage E_L^+ and a reflected voltage E_L^-.

$$E_L = E_L^+ + E_L^- \tag{2.16}$$

The size of the reflection is called the *reflection coefficient* and is denoted by Γ.

$$\Gamma = \frac{E_L^-}{E_L^+} = \frac{Z_L - Z_0}{Z_L + Z_0} \tag{2.17}$$

It is customary to consider all impedances as multiples of the characteristic impedance and to use lower case letters to designate these "normalized" impedances. In Eq. (2.17) both numerator and denominator of the right-hand side should be divided by Z_0 to normalize the impedances.

$$\Gamma = \frac{(Z_L/Z_0) - 1}{(Z_L/Z_0) + 1} = \frac{z_L - 1}{z_L + 1} \tag{2.18}$$

The ratio of reflected current to incident current is the negative of the reflection coefficient:

$$\frac{I_L^-}{I_L^+} = -\frac{E_L^-}{E_L^+} = -\Gamma \tag{2.19}$$

The reflection coefficient, then, always applies to voltages. In general, it is complex, since the reflected voltage may have any phase relationship with the incident voltage. As indicated in Eq. (2.19), the phase difference between E_L^+ and I_L^+ is 180° different from that between E_L^- and I_L^-.

Obviously, the reflection coefficient can have any value from zero, where nothing is reflected, to unity, where there is total reflection. If the load impedance is equal to the characteristic impedance, $Z_L = Z_0$, then from Eq. (2.17), $\Gamma = 0$. This means there is no reflection, and all power goes into the load. If the line is short-circuited, $Z_L = 0$. Now from Eq. (2.17), $\Gamma = -1$; that is, all the power is reflected, and the reflected voltage at the load is 180° out of phase with the incident voltage there —$E_L^- = -E_L^+$. The actual voltage across the load is the sum of the incident and reflected voltages (indicated in Eq. (2.16)) and is therefore zero. This is to be expected since there can be no voltage across a short circuit. If the line is open-circuited, $Z_L = \infty$; Eq. (2.17) becomes

$$\Gamma = \frac{\infty - Z_0}{\infty + Z_0} = \frac{1 - (Z_0/\infty)}{1 + (Z_0/\infty)} = 1 \tag{2.20}$$

This again indicates total reflection, but in this case, the incident and reflected voltages are in phase. The total voltage at the open circuit is now twice the incident voltage. The reflected current has a phase reversal as indicated in Eq. (2.19). Thus the total current is zero, which is to be expected since no current can flow across an open circuit. These three cases, matched load, open circuit, and short circuit are very important.

If the load impedance Z_L is a pure reactance of the form jX, Eq. (2.17) becomes

$$\Gamma = \frac{jX - Z_0}{jX + Z_0} \tag{2.21}$$

The magnitude of Γ is then

$$|\Gamma| = \frac{\sqrt{X^2 + Z_0^2}}{\sqrt{X^2 + Z_0^2}} = 1 \tag{2.22}$$

which indicates total reflection. In other words, a pure reactance cannot absorb power. The phase of the reflection may have any value from 0 to 2π radians depending on the size and sign of the load reactance.

2.8. STANDING WAVES

When there is a reflection from a discontinuity or from the end of a transmission line, part or all of the incident wave is made to travel back toward the input end. Thus, between the input end and the source of the reflection there will be two waves traveling in opposite directions. At some points in the line the two waves will always be in phase and will add, while at other points the two will always be out of phase and will cancel. The places where the two waves add will be points of maximum voltage, while the points of cancellation will have minimum voltage. In the case of complete reflection, where $\Gamma = 1$, the cancellation is complete and the voltage minimum is zero. Since the positions of maximum and minimum voltage remain motionless, a *standing wave* is said to exist on the line.

2.9. TOTAL REFLECTION

The simplest standing waves occur when the reflection coefficient Γ is unity. As has been mentioned, this occurs when the load impedance is a short circuit, an open circuit, or a pure reactance. It does not occur if the load has a resistive component which will absorb some of the incident power.

If the load is an open circuit, Eq. (2.20) shows that the reflected voltage and incident voltage are the same and therefore add at the termination or load. Thus the standing wave has a voltage maximum at the open-circuited end. A quarter-wavelength from the end, the incident wave will be 90° earlier and the reflected wave 90° later than they are at the end, and will thus be 180° out of phase. At this point the voltage will be zero.

Similar reasoning shows that the standing wave pattern is repeated every half-wavelength; that is, maxima are spaced half a wavelength apart on the transmission line and minima are also spaced half a wavelength apart. The distance between a maximum and a minimum is a quarter-wavelength. The current and voltage are in phase at a maximum or a minimum; at all other points, assuming there is a standing wave, the current and voltage are out of phase. The current maximum occurs at a point of voltage minimum and vice versa. All the statements in this paragraph apply not only to standing waves resulting from open circuits but to all standing waves.

If the load is a short circuit, the current is maximum at the termination while the voltage is zero. The standing wave thus has a node or minimum at the end and at every half-wavelength from the end. It should be noted that the term "standing wave" refers to a voltage standing wave unless the word "current" or "power" explicitly precedes it.

If the load is a pure reactance, neither the maximum nor the minimum of the standing wave occurs at the end of the line. However, the standing wave behaves as it does in a short or open circuit except it is shifted along the line. Figure 2.6 shows standing waves resulting from loads causing

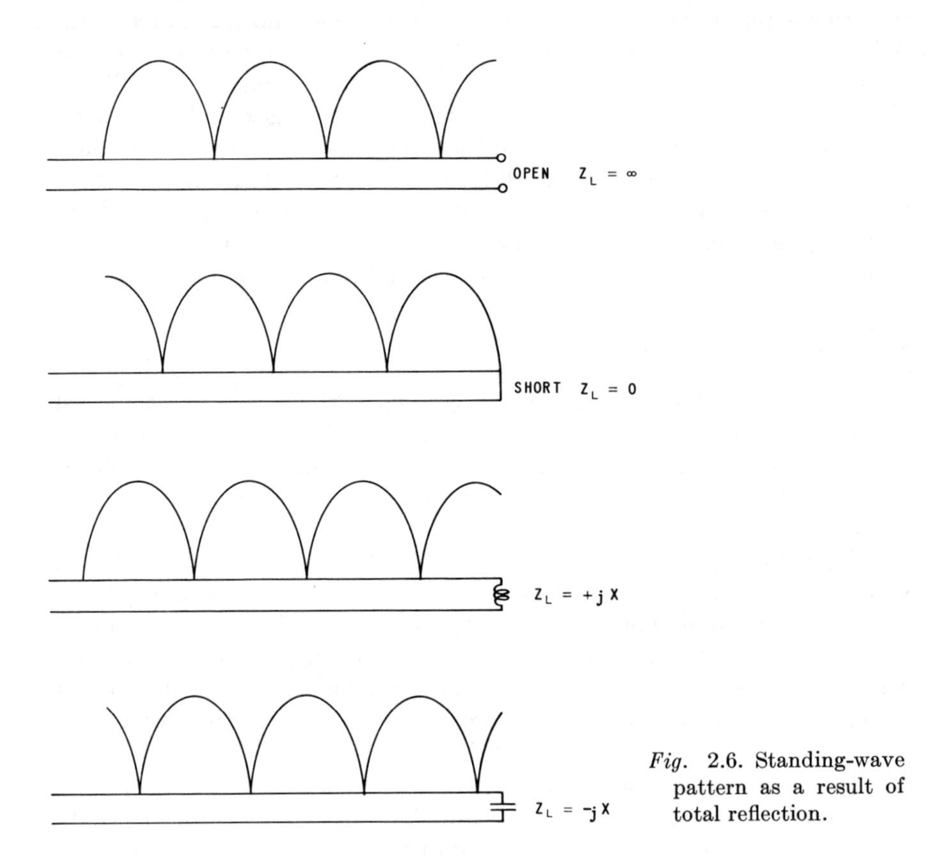

Fig. 2.6. Standing-wave pattern as a result of total reflection.

total reflections. In each case the voltage at maximum is twice the incident voltage, and the voltage at minimum is zero.

2.10. PARTIAL REFLECTION

From Eq. (2.17) it may be seen that if Z_L is real and not equal to Z_0, Γ will be real and will have an absolute value of less than unity. This means that some of the power will be absorbed by the load and only part of it will be reflected. From Eq. (2.18) it may be observed that if Z_L is twice Z_0 (that is, if $z_L = 2$), $\Gamma = \frac{1}{3}$. Also if Z_L is half of Z_0, $\Gamma = -\frac{1}{3}$. In both cases the reflected voltage is $\frac{1}{3}$ of the incident, but in the former case the reflected will be added to the incident voltage at the load, while in the latter it will be subtracted. Stated otherwise, when $Z_L = 2Z_0$, the voltage at the load is $1\frac{1}{3}$ times the incident voltage. When $Z_L = Z_0/2$, the voltage is $\frac{2}{3}$ of the incident voltage. In both cases the standing wave voltage will vary from $1\frac{1}{3}$ times at a maximum to $\frac{2}{3}$ at a minimum; a maximum will be at the load when Z_L is greater than Z_0, and a minimum will be there when Z_L is less than Z_0.

The considerations that determine the position of the maximum and its amplitude are the same when Z_L is complex as when the loads are simple.

2.11. STANDING-WAVE RATIO

On a lossless line, the voltage maxima of a standing wave all have the same amplitude. Likewise, all the minima have the same amplitude. For practical cases in short lines, it may be assumed that there is no loss. The ratio of the maximum voltage on the line to the voltage at a minimum is called the *standing-wave ratio* or, more exactly, the *voltage standing-wave ratio*. To avoid ambiguity it is usually abbreviated VSWR. Thus

$$\text{VSWR} = \frac{E_{\text{max}}}{E_{\text{min}}} \tag{2.23}$$

The maximum voltage is the sum of the incident and reflected voltages, while the minimum voltage is the difference. Since the reflected voltage is proportional to the absolute magnitude of the reflection coefficient, the standing-wave ratio becomes

$$\text{VSWR} = \frac{E_{\text{max}}}{E_{\text{min}}} = \frac{1 + |\Gamma|}{1 - |\Gamma|} \tag{2.24}$$

Solving for $|\Gamma|$, this becomes

$$|\Gamma| = \frac{\text{VSWR} - 1}{\text{VSWR} + 1} \tag{2.25}$$

It should be noted that $|\Gamma|$ must lie between zero and one. If $\Gamma = 0$, there is no reflection; if $|\Gamma| = 1$, there is total reflection. If $|\Gamma|$ were greater than unity, it would indicate that the reflected signal was greater than the incident signal. The VSWR is never less than unity, and it may have any value from one to infinity. It is infinite when there is total reflection; it is unity when there is no reflection.

The value of the standing-wave ratio is determined by the reflection coefficient, as indicated in Eq. (2.24). This and the position of the voltage minimum nearest the load are sufficient to determine the load impedance. The standing-wave ratio may be measured by simply testing the voltage along a transmission line. At microwaves, where lines are usually shielded, a longitudinal slot is provided in the wall of a coaxial line or waveguide; a small probe is inserted to sample the voltage and is slid along the line to find the position of the minimum, as well as the ratio of the maximum voltage to the minimum voltage. The piece of transmission line with the slot in it is called a *slotted section* or a *slotted line.*

The slot in a slotted section must be cut in such a way that no appreciable power leaks out of the line. In a coaxial line, the slot is cut in the outer conductor, parallel to the direction of propagation. Thus, it is parallel to the flow of current in the outer conductor and causes negligible interference with the current. In a waveguide, the slot is cut in the center of a broad wall, and again it is parallel to the direction of propagation for similar reasons.

The inserted probe acts as an antenna to receive a small portion of the signal at each point in the slotted line. An indicating instrument is connected to the probe to allow the amplitude of the voltage to be read. Most of the detectors used in the laboratory for this purpose are so-called *square-law* detectors, receiving that name because the output d-c voltage is proportional to the square of the voltage at the probe. If the detector is connected to an ordinary voltmeter, the readings will be proportional to power or will give a standing-wave power ratio. The VSWR is obtained by taking the square root of this power ratio; however, most laboratories usually use special meters for reading VSWR directly with a square-law detector.

When a transmission line is used to carry power from a source to a load, it is obvious that it is desirable to have all the power transferred to the load. If any power is reflected, it does not reach the load and thus is wasted. The VSWR is an indication of the amount of power reflected. A perfect line would have a VSWR of unity, but this is usually difficult to obtain over a wide frequency range. For many applications a VSWR of 2 is satisfactory, but it is sometimes necessary to attain values as low as 1.1, or even less in special applications.

2.12. POWER

In any circuit the instantaneous power is given by the product of the voltage and current at that instant. Thus, in a d-c circuit $P = EI$, but in an a-c circuit, the voltage and current may be out of phase. In a transmission line, a simple way to determine power is to measure voltage and current at a point where they are in phase. A suitable point is a voltage maximum. At this point the current is minimum but in phase with the voltage.

$$P = E_{\max} I_{\min} \tag{2.26}$$

$$I_{\min} = |I^+| - |I^-| = \frac{|E^+|}{Z_0} - \frac{|E^-|}{Z_0} = \frac{E_{\min}}{Z_0} \tag{2.27}$$

Therefore,

$$P = \frac{E_{\max} E_{\min}}{Z_0} \tag{2.28}$$

If there is no standing wave, as in a perfectly matched line, then

$$E_{\max} = E_{\min} = |E^+|$$

$$P = \frac{|E|^2}{Z_0} \tag{2.29}$$

2.13. IMPEDANCE

The *impedance* Z at any point in a line is the ratio of the voltage to the current at that point. Since the voltage and current are not necessarily in phase, Z is generally complex.

$$Z = R + jX \tag{2.30}$$

where the real part R is called *resistance,* and the imaginary part X is called *reactance.* If X is positive, the reactance is inductive. Negative reactance is capacitive.

The reciprocal of impedance is called *admittance* and is designated Y. It is obviously the ratio of current to voltage and is also complex.

$$Y = G + jB \tag{2.31}$$

where G is real and is known as *conductance,* and B is imaginary and is known as *susceptance.* Inductive susceptance as defined here is negative, and capacitive susceptance is positive.

The voltage and current at a point in a line are dependent on the load at the end of the line and on the distance of the point from the load. The impedance then must also be dependent on the load and the distance to it. If the unknown point is at the input end of the line, the input impedance is a function of the length of the line and the load at the end.

The impedance at a distance s from a load is

$$Z_s = \frac{E_s}{I_s} = \frac{E_L \cos \beta s + jZ_0 I_L \sin \beta s}{I_L \cos \beta s + j \dfrac{E_L}{Z_0} \sin \beta s} \tag{2.32}$$

where E_s and I_s are at the point which is a distance s from the load, E_L

and I_L are at the load, and β is the phase constant. βs is thus the electrical distance from the load. If the numerator and denominator of the right-hand side of Eq. (2.32) are multiplied by $Z_0/I_L \cos \beta s$

$$Z_s = Z_0 \frac{Z_L + jZ_0 \tan \beta s}{Z_0 + jZ_L \tan \beta s} \tag{2.33}$$

Of course, if s is the length of the line, Z_s is the input impedance. It should be noted that $\tan \beta s$ is a periodic function which recurs every 180°, or half-wavelength, as βs is increased.

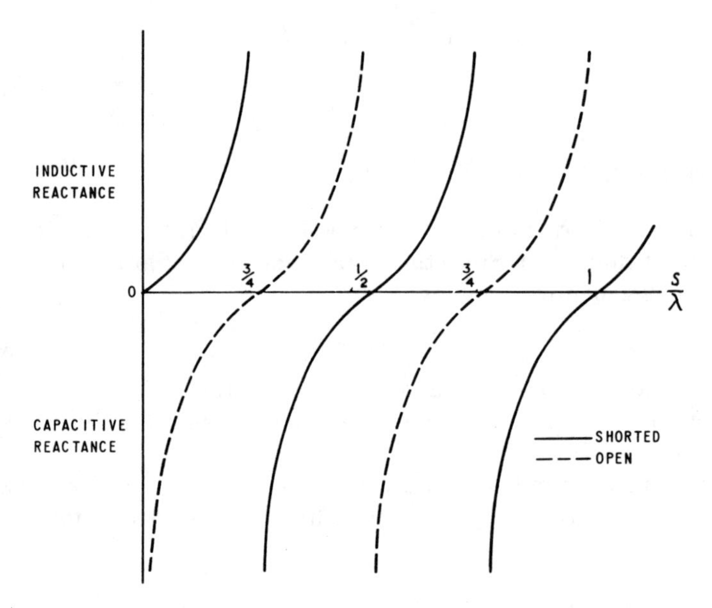

Fig. 2.7. Reactance of shorted and open-circuited lines.

Equation (2.33) is very important. From it, it is possible to determine the input impedance with any load, or to pick a load or a line to arrive at specified input conditions. There are several special cases which are of major interest.

If a line is terminated in its characteristic impedance, that is if $Z_L = Z_0$, then from Eq. (2.33),

$$Z_s = Z_0 \tag{2.34}$$

This is true for any value of s and any value of β. If s is fixed, β may be varied by changing frequency.

If a line is terminated in a short circuit, that is, if $Z_L = 0$,

$$Z_s = jZ_0 \tan \beta s \tag{2.35}$$

If a line is terminated in an open circuit, $Z_L = \infty$,

$$Z_s = -jZ_0 \cot \beta s \tag{2.36}$$

Equations (2.35) and (2.36) both indicate that the input impedance is a pure reactance. (This is plotted in Fig. 2.7.) The similarity of these impedances to those of resonant circuits may be seen by comparing Fig. 2.7 to Fig. 2.8, which shows the reactances of a series resonant circuit and of a shunt resonant circuit. The lumped circuits are resonant at single

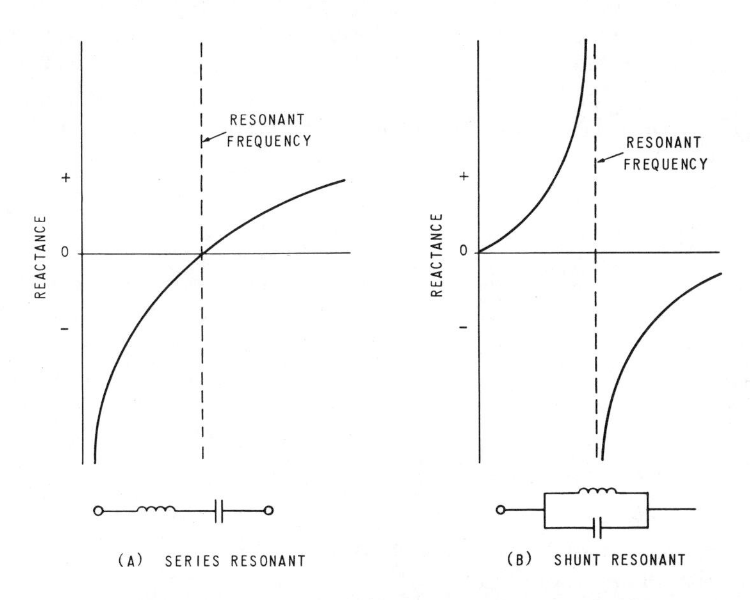

Fig. 2.8. Reactance vs. frequency for resonant circuits.

frequencies, while the shorted and open lines are resonant at many frequencies. The shorted line presents infinite impedance similar to a shunt resonant circuit when the length is any odd multiple of a quarter-wavelength. The open-circuited line presents zero impedance similar to a series resonant circuit at the same lengths. At multiples of a half-wavelength, the shorted line behaves as a series resonant lumped circuit and the open line appears to be shunt resonant.

When a line is a half-wavelength long or any multiple thereof, $\tan \beta s = 0$ and $Z_s = Z_L$.

When a line is a quarter-wavelength long or an odd multiple of a quarter-wavelength long, $\tan \beta s = \infty$. Then dividing numerator and denominator by $\tan \beta s$,

$$Z_s = Z_0 \frac{(Z_L/\infty) + jZ_0}{(Z_0/\infty) + jZ_L} = \frac{Z_0^2}{Z_L} \tag{2.37}$$

or

$$Z_s Z_L = Z_0^2$$

It should be noted that if Z_L is high, Z_s will be low and vice versa. Because a quarter-wavelength line has this ability to transform an impedance, it has been called a *quarter-wavelength transformer*. From Eqs. (2.35) and (2.36) or from Eq. (2.37), it can be seen that a quarter-wavelength of line transforms an open circuit to a short circuit and a short circuit to an open.

Sometimes it is more important to know the input admittance rather than the input impedance. Of course, one is the reciprocal of the other. Thus, if the impedance is found by Eq. (2.33), the admittance is

$$Y_s = \frac{1}{Z_s} \tag{2.38}$$

However, if instead of Z_0 and Z_L, corresponding admittances are known, it would be awkward to change all the admittances to impedances, find Z_s, and change back to Y_s. Fortunately, the relationships among admittances are the same as among impedances, so that Y_s may be found directly as

$$Y_s = Y_0 \frac{Y_L + jY_0 \tan \beta s}{Y_0 + jY_L \tan \beta s} \tag{2.39}$$

Both Eqs. (2.33) and (2.39) are usually normalized by assuming Y_0 and Z_0 are unity. They then become

$$\begin{aligned} z_s &= \frac{z_L + j \tan \beta s}{1 + jz_L \tan \beta s} \\ y_s &= \frac{y_L + j \tan \beta s}{1 + jy_L \tan \beta s} \end{aligned} \tag{2.40}$$

2.14. EXAMPLES

Equations (2.33) and (2.39) or (2.40) may be used to solve many problems involving impedances.

Example 2.1. To find the input impedance of a line. Given a line of characteristic impedance 100 ohms and a third of a wavelength long. If this line is terminated in a load which has an impedance $150 + j60$, what is its input impedance? In this problem, $Z_0 = 100$, $Z_L = 150 + j60$, $\beta s = 360°/3 = 120°$.

Then $\tan \beta s = -\sqrt{3}/2$. From Eq. (2.33),

$$Z_s = 100 \frac{150 + j60 + j100(-\sqrt{3}/2)}{100 + j(150 + j60)(-\sqrt{3}/2)}$$

which is solved by simple arithmetic.

Example 2.2. To find the load impedance. Given a line of characteristic impedance 100 ohms and an eighth of a wavelength long. If the input impedance is $50 + j7$, what is the load impedance? In this problem, $Z_0 = 100$, $\beta s = 360°/8 = 45°$, $Z_s = 50 + j7$. Then $\tan \beta s = 1$. From Eq. (2.33),

$$50 + j7 = 100 \frac{Z_L + j100}{100 + jZ_L}$$

which is solved for Z_L by simple algebra. Z_L will be complex of the form $R_L + jX_L$.

In simple cases, impedance problems may be solved as indicated. If special conditions exist so that one of Eqs. (2.35), (2.36), or (2.37) is applicable, the problem is usually solved by using the appropriate equation. However, in more complicated examples, including perhaps the two mentioned above, it is more convenient to solve the problem by using some form of impedance chart.

2.15. THE SMITH CHART

Figure 2.9 illustrates the Smith Chart, an impedance chart devised by P. H. Smith.* As pictured here, the diameter marked "Resistance Component" is horizontal. This agrees with Smith's original presentation, although many engineers use the chart with this line vertical.

Referring to Fig. 2.9 it is evident that the chart contains two sets of lines. The lines of constant resistance form circles, all tangent to each other at the right-hand end of the horizontal diameter. The value of resistance along any of these circles is indicated above the diameter and does not change. These values start at zero at the left and proceed to infinity at the right. The center is labeled 1.0. It must be remembered that this is a normalized chart and that the center 1.0 means Z_0. Thus, if Z_0 is 50 ohms, the circle that passes through the center represents 50 ohms, the circle that passes through 0.2 represents ten ohms, the circle that passes through 3.0 represents 150 ohms, etc.

The second set of lines represent constant reactance. These lines are all arcs of circles which are also (like the circles of constant resistance) tangent to each other at the right-hand end of the horizontal diameter, and also tangent to this diameter. Values of reactance in the upper half of

* P. H. Smith, "Transmission Line Calculator," *Electronics*, Jan. 1939.

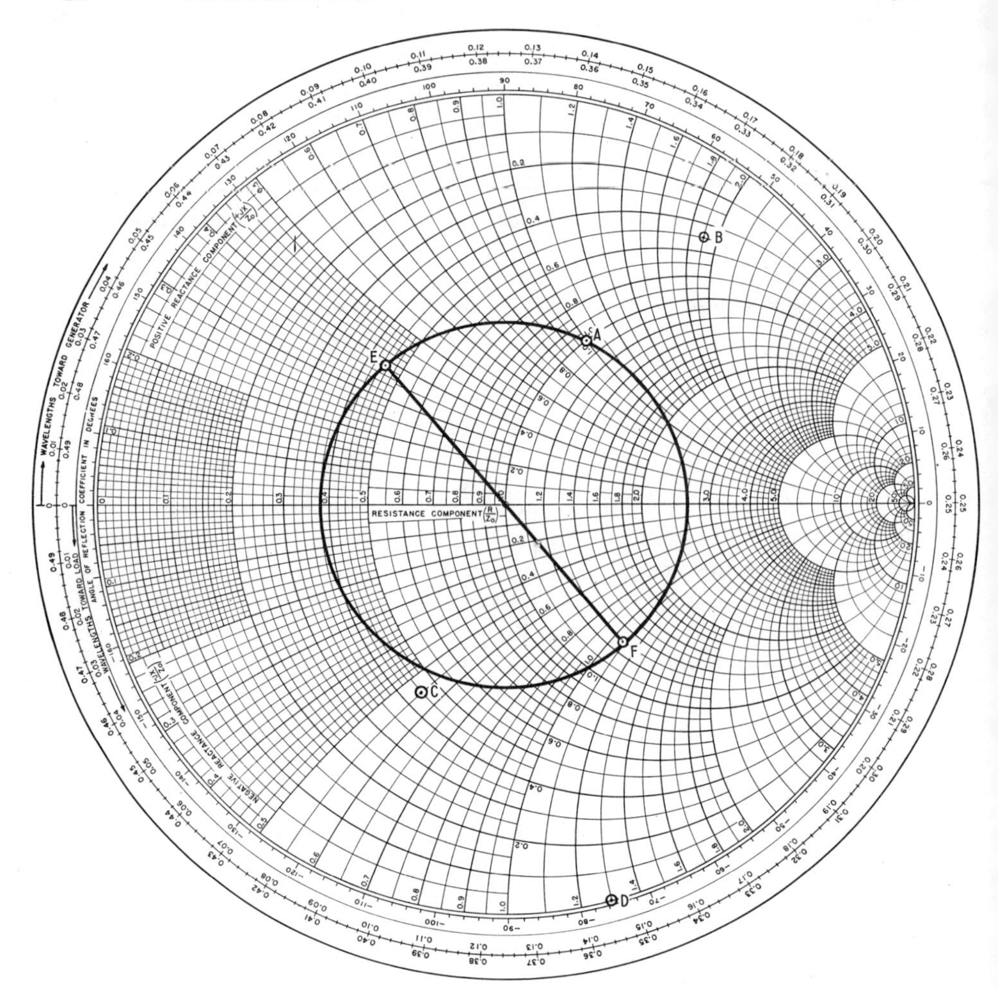

Fig. 2.9. Smith Chart.

the diagram are positive while those below are negative. This information is usually located on the chart to prevent errors. As with the resistance values, the reactances are normalized and must be multiplied by Z_0 to get values in ohms.

Reading the impedance from the chart is quite simple. The center is, of course, unity; that is, the normalized impedance is $1 + j0$. Several points are shown to illustrate the reading of impedances. Point A is $1 + j1$. Point B is $0.5 + j1.9$. Point C is $0.45 - j0.55$. Point D is $0 - j1.3$. The left-hand end of the horizontal diameter is zero and the right-hand end infinity.

The quantity Γ, the reflection coefficient, doesn't appear on the chart, but, nevertheless, it is implied. The outermost circle, passing through the zero and infinity points, has a radius which is considered unity for Γ. The distance of any point from the center is then a measure of its reflection coefficient. So point C is about half way between the center and circumference, and thus has a reflection coefficient of about one half.

The impedance at any point in a line looking toward the load has been shown to be dependent on the load at the end and on the distance to that load. Equation (2.33) indicates that as the point is moved along the line, this impedance changes and repeats every half-wavelength. At the points of maximum and minimum voltage, this impedance is respectively maximum or minimum, and is real. The VSWR looking into the line is Z_{max}/Z_0 or Z_0/Z_{min}. This means that a point on the chart representing the impedance at a point in a line also represents the impedance at a point half a wavelength away from the original point. As the point in the line is moved along the line, assuming a lossless line of constant characteristic impedance Z_0, the point on the chart will move in a circle with its center at the center of the chart. If the point in the line moves toward the generator, the point on the chart moves clockwise, as indicated by an arrow on the outside of the chart. Movement toward the load is counterclockwise. When the point crosses the real axis between 1.0 and ∞, the impedance is real (resistive). The point of crossing also gives the VSWR since it is Z_{max}/Z_0.

The circle through A in Fig. 2.9 passes through all values of impedance that can exist in a line of constant Z_0 if the point A is one value of impedance. This circle crosses the real axis at the right at 2.6, which means that the VSWR is 2.6. From Eq. (2.25), the reflection coefficient corresponding to this VSWR is $\Gamma \doteq 0.44$. This means that the points on the circle through A should be about 0.44 of the way from the center of the big circle to the circumference.

Since Eq. (2.39) is of the same form as Eq. (2.33), the Smith Chart may be used for admittances as well as impedances. Thus if Fig. 2.9 is an admittance chart (instead of an impedance chart), point A represents an admittance $Y = G + jB = 1 + j1$. Again, this is normalized and must be multiplied by the value of Y_0 to get the true value.

Using the Smith Chart it is very simple to change from impedance to admittance. The value of admittance corresponding to any impedance is always diagonally opposite and the same distance from the center. Thus, the point E in Fig. 2.9 may represent an impedance $0.45 + j0.38$. The value of admittance corresponding to this is

$$\frac{1}{0.45 + j0.38} = 1.3 - j1.08$$

As may be seen, this is the value at the point F which is the same distance from the center as E and diagonally opposite.

2.16. SMITH CHART SOLUTIONS TO PROBLEMS

The examples of Sec. 2.14 may be solved more simply by using Smith Charts than by using the equations. Repeating Ex. 2.1: "Find the input impedance of a line. Given a line of characteristic impedance 100 ohms and a third of a wavelength long. If this line is terminated in a load which has an impedance $150 + j60$, what is its input impedance?" Figure 2.10 shows the solution. First, everything is normalized to

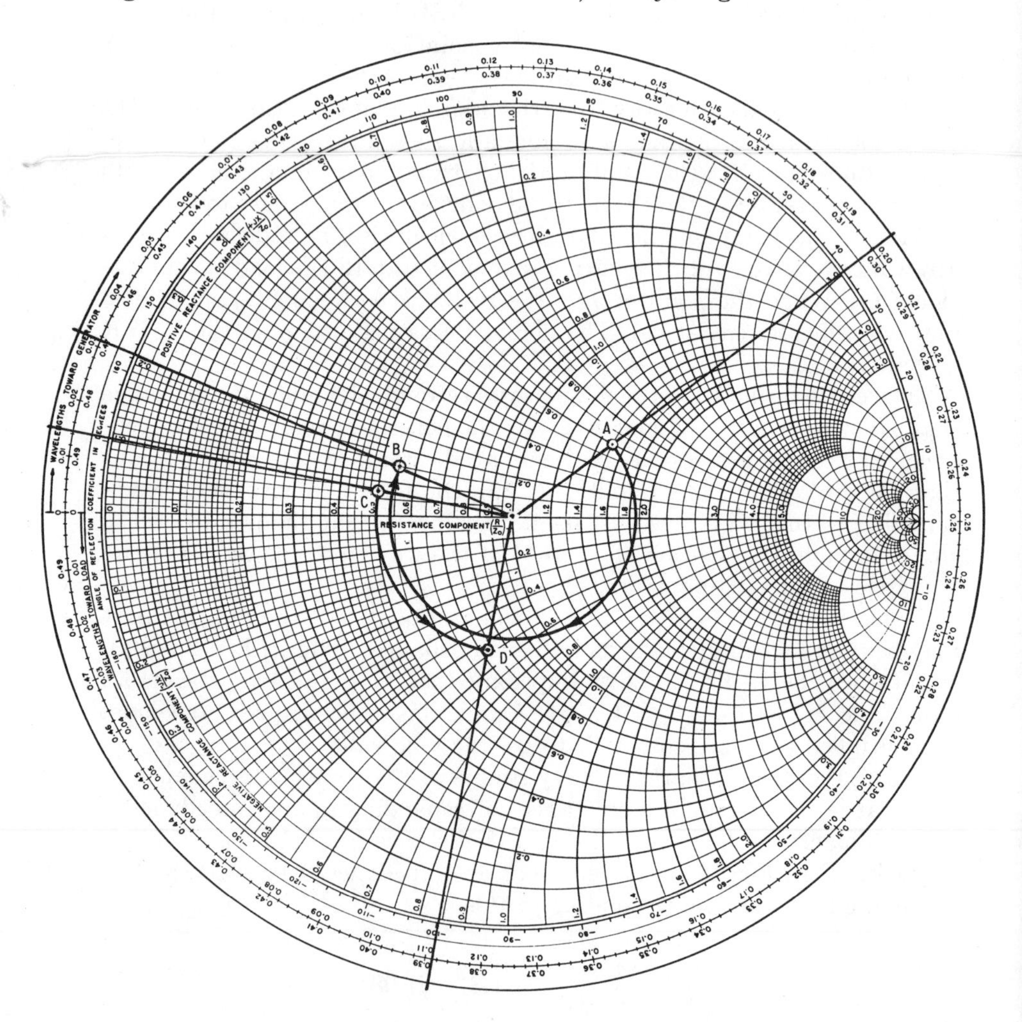

Fig. 2.10. Problem solution using Smith Chart.

$Z_0 = 100$. Thus, $z_L = 1.5 + j0.6$. This point is located on the chart at point A, and a radius is drawn through it. The radius intercepts the "wavelengths toward generator" circle at 0.198. As the impedance is moved a third of a wavelength from the load toward the generator, point A on the chart describes an arc of a circle clockwise. Since the line's characteristic impedance is constant at Z_0, the center of the circle described by A lies at the center of the Smith Chart, and corresponds to Z_0. The new radius one-third of a wavelength away should pass through 0.531 (= 0.198 + 0.333). The Smith Chart, like the transmission line, is periodic in $\lambda/2$. Therefore 0.531λ on the chart is the same as 0.031λ. Point A has rotated one-third wavelength to B, which is $0.55 + j0.15$. Multiplying by the value of Z_0, the input impedance is found to be $55 + j15$ ohms.

Example 2.2 is treated similarly. Repeating Ex. 2.2: "Find the load impedance. Given a line of characteristic impedance 100 ohms and an eighth of a wavelength long. If the input impedance is $50 + j7$, what is the load impedance?" The input impedance is located on the chart at point C. The radius passes through 0.486 on the "wavelengths toward load" circle. An eighth of a wavelength toward the load moves the radius to 0.611 (= 0.486 + 0.125) which is the same as 0.111 on the circle. Point D now indicates the load impedance as $0.73 - j0.53$. Multiplying by Z_0, the load is $73 - j53$ ohms.

Of course, if a line is terminated in a short circuit or an open circuit, it would be easier to use Eq. (2.35) or (2.36) than to plot the problem on a Smith Chart. For example, if a short-circuited line is 60° long ($\chi/6$) with a characteristic impedance of 50 ohms, the input impedance from Eq. (2.35) is

$$Z_s = j50 \times 1.732 = j86.6 \text{ ohms}$$

This is a pure reactance.

2.17. REACTIVE STUBS

As has been mentioned, short-circuited and open-circuited lines behave as pure reactances. Consequently, they absorb no power from the circuit; in fact they may be used to assist in transmitting all the power to its destination. A simple example is that of delivering power to an antenna which has a complex impedance. The antenna impedance is located on a Smith Chart and rotated to a point where it is on the $R = 1$ line. In general, its impedance will be $1 + jX_a$. If at this point in the transmission line a series reactance $-jX_a$ is inserted, the total impedance will be

$$1 + jX_a - jX_a = 1$$

and there will be no reflection. Such a series reactance could be a short-circuited length of line, usually called a *shorted stub*. In practice, an open-circuited line is rarely used at microwaves, since a true open circuit is virtually impossible to obtain. If a line is left open, it is really terminated in the impedance of free space. At the termination there will be some loss due to radiation.

If admittances are used instead of impedances, the stub may be shunted across the line. In this case, the admittance is rotated to a point where it is $1 + jB_a$. Then a stub is put in shunt with a susceptance of $-jB_a$, making the total admittance

$$1 + jB_a - jB_a = 1$$

and, as before, there will be no reflections. X_a and B_a in these examples can be either positive or negative; the shunt stub is used more often than

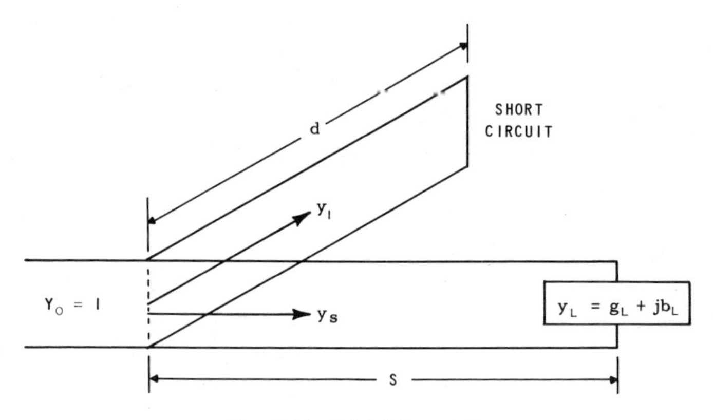

Fig. 2.11. Matching stub.

the series stub since it can be added to a line without cutting it. Figure 2.11 shows schematically how a shunt stub is used. The load is complex, $g_L + jb_L$. By moving along the line (to the left in the figure) a distance s, a point is found where $y_s = 1 + jb_s$. A short-circuited stub is placed in shunt with the line at this point, and its length d is adjusted until the admittance y_l looking into it is $-jb_s$. The total admittance is then $y_s + y_l = 1$. In practice, the position of the stub is determined by using a chart, but the short circuit is adjustable. Thus, the distance d may be found empirically by moving the shorting plunger until the reflection is minimized.

2.18. DOUBLE-STUB TUNER

In using the single stub, it was necessary to locate the point in the line where this stub would operate correctly. In many transmission lines, such as coaxial lines or waveguides, it is inconvenient to put the stub in its place after that place is located. However, two stubs, suitably positioned, can be fixed in place and will still be able to cancel out most reflections. This device is called a *double-stub tuner*. It consists of two stubs, fixed in position, but with adjustable shorting plungers. The spacing of the stubs is usually three-eighths of a wavelength, but may be closer.

Matching with a double-stub tuner may be calculated, but it is usually simpler to do it by trial and error. The stubs are adjusted alternately until the VSWR is reduced to unity. In some cases, the conductance of the load may be too high at the first stub and complete matching may not be possible. In such a case, a length of line about a quarter-wavelength long may be inserted between the load and the double-stub tuner. A quarter-wavelength line changes a high value of conductance to a low one.

2.19. BALUNS

At lower radio frequencies there are two kinds of transmission lines in general use, the open wire line and the coaxial line. The open wire line may be a pair of parallel wires or a twisted pair, but in either case both

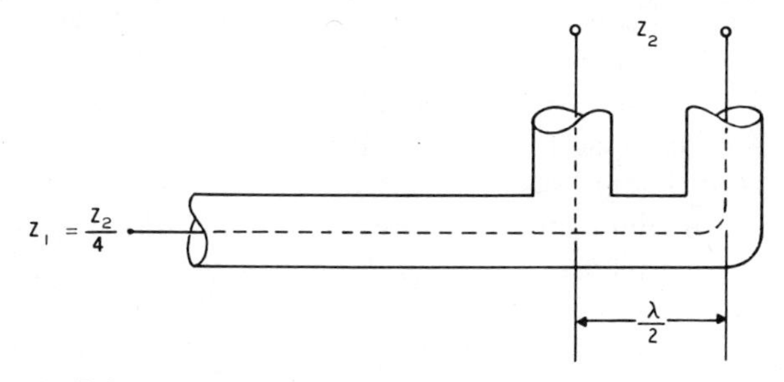

Fig. 2.12. Balun.

wires have the same relationship to ground. Such a line is called a *balanced* line. In the coaxial line the outer conductor is usually grounded so that the two conductors do not have the same relationship to ground. This is, therefore, an *unbalanced* line.

It is sometimes necessary to change from a balanced to an unbalanced line, and this is accomplished by a balanced-to-unbalanced transformer, which is usually called a *balun*. It is obvious that the balanced line conductors cannot be connected directly to the conductors of the unbalanced line, for the one connected to the outer conductor of the coaxial line would be grounded. A simple balun is shown in Fig. 2.12. The inner conductor of the coaxial line is tapped off 180°, or half a wavelength, from the end. The tap and the extension of the inner conductor at the end make a balanced line since the voltages on each half are 180° out of phase and have identical relationships to ground. The impedance of the balanced line should be four times the impedance of the coaxial line since the voltages at the balanced line add in series while at the coaxial end they are in parallel. This type of balun is thus useful for connecting a 300-ohm open-wire line to a 75-ohm coaxial line. Its major limitation is frequency sensitivity; the length of line between the two parts of the balanced output is a half-wavelength long at only one frequency.

This kind of circuit element is not really a microwave component. However, at microwaves it is sometimes necessary to borrow techniques from lower frequencies. Some microwave systems use a half-wave dipole as a radiator. This type of antenna is inherently balanced. Consequently, if a coaxial line is used to feed it, some form of balun must be used. The simple balun shown in Fig. 2.12 is the basic circuit, but more sophisticated components have less frequency sensitivity and can match impedances with ratios other than four to one.

2.20. MULTIPLE REFLECTIONS

When two or more discontinuities in a transmission line send reflections back toward the source, these reflections may have any phase relationship with each other. If the reflections are all in phase, they will add and the total reflection will be greatest. However, it is possible for the phase relationships to cause the total reflection to be zero. In this case, if all the discontinuities are lossless, there will be no power lost in the circuit, and all the incident power will reach the load. As will be seen in Sec. 6.5, this is a frequently used matching device. A discontinuity is purposely added to a circuit in such a manner that its reflection is equal to and out of phase with, the reflection already in the circuit. The net reflection is then zero.

Between two discontinuities, part of the signal bounces back and forth many times. Lines are not lossless, and consequently the more bouncing, the greater the attenuation of the signal. However, for the small discontinuities most often encountered in normal microwave circuits, this loss is usually neglected since it is minute compared to the over-all loss of the system.

If two components, each of which has a known VSWR, are to be combined in series in a circuit, it is possible to calculate the total VSWR of the circuit. In practice, it is usually sufficient to know the maximum VSWR that will be encountered. If the two VSWR's are designated r_1 and r_2, the maximum possible VSWR is the product of the two:

$$r_{\text{max}} = r_1 r_2 \tag{2.41}$$

The minimum VSWR may also be determined:

$$r_{\min} = \frac{r_1}{r_2} \qquad (r_1 \geq r_2) \tag{2.42}$$

In some circuits, it is possible to measure the maximum and minimum VSWR by sliding one discontinuity along the line. (This is frequently done with absorbing loads.) The two VSWR's, r_1 and r_2, may then be found by solving Eqs. (2.41) and (2.42) as

$$r_1 = \sqrt{r_{\max} r_{\min}} \tag{2.43}$$

$$r_2 = \sqrt{\frac{r_{\max}}{r_{\min}}} \tag{2.44}$$

In the case of a sliding load, one of these is the VSWR of the load and the other is the VSWR of the rest of the circuit.

QUESTIONS AND PROBLEMS

2.1. List three functions of a transmission line.

2.2. What factors determine the proper choice of line or guide for a given transmission problem?

2.3. Discuss briefly three types of transmission line energy losses.

2.4. Define and explain the significance of the term "skin depth."

2.5. What is the approximate skin depth on the copper-center conductor of a coaxial line transmitting a signal at:
a. 100 Mc?
b. 500 Mc? *ans.* a. 0.00026 in.

2.6. A transmission line with a surge impedance of 500 ohms is terminated with a 100-ohm load. One hundred watts of energy (continuous sine-wave signal) is being dissipated in the load.
a. What is the reflection coefficient of the load?
b. What is the magnitude of the reflected voltage? *ans.* 200 v
c. What is the VSWR on the line?
d. What is the highest voltage the line dielectric must withstand under the conditions given? *ans.* 707 v

2.7. The voltage variations along a 50-ohm line are checked with a probe. The voltage is found to be 20 v at a loop and at a node 5 v.
a. What is the VSWR on the line?
b. What is the magnitude of the reflection coefficient?
c. Find the power being dissipated in the load.

2.8. Find the input impedance of a quarter-wave section of a 50-ohm line when it is terminated with a 200-ohm load. Repeat the problem with a terminating load of 10 ohms. Discuss the statement, "a quarter-wave line inverts the load."

2.9. Discuss the following statements:
a. A $\frac{1}{2}$ wave transformer repeats the load.
b. A $\frac{1}{2}$ wave transformer acts like a 1:1 transformer.

2.10. Find the input impedance of a one-eighth-wavelength section of a 50-ohm transmission line terminated with:
a. 25 ohms resistance.
b. 50 ohms inductive reactance.
c. $10 + j20$ ohms. *ans.* a. $40 + j30$

2.11. Use a Smith Chart to find the parallel circuit equivalent of the series impedance $50 + j40$ ohms.

2.12. A piece of 50-ohm line one sixth of a wavelength long is terminated with a 25-ohm resistor. Find the input impedance of this line using Eq. (2.33) and check the result by means of the Smith Chart.

2.13. What is the magnitude of the reflection coefficient if:
a. A 50-ohm line is terminated with $50 - j20$ ohms?
b. A 50-ohm line is terminated with $20 + j50$ ohms? *ans.* a. 19

2.14. You wish to determine the value of the load at the end of a section of 100-ohm line. The VSWR on the line is checked and found to be 2.5. A voltage node is found 0.1 wavelength from the load. What is the value of the terminating impedance? *ans.* $58 - j57$

2.15. What is the advantage of double-stub over single-stub matching?

2.16. A slotted line is placed in the line leading to a load. The manufacturer of the slotted section gives its maximum VSWR as 1.1. The VSWR of the load is found to be 1.3. What is the maximum value of VSWR that can exist ahead of the slotted section?

3

MICROWAVE MEASUREMENTS

At low frequencies, it is convenient to measure voltage and current and use them to calculate power. However, at microwave frequencies, they are difficult to measure and, since they vary with position in a transmission line, are of little value in determining power. Therefore, at microwaves it is more desirable and, in fact, simpler to measure power directly.

Again, at low frequencies, circuits use lumped elements which can be identified and measured. At microwaves, where distributed circuits or distributed elements are used, it is usually not important to know what elements could make up a line. It is possible and satisfactory to measure the impedance of a circuit without regard to the individual distributed elements making up that circuit.

Unlike low-frequency measurements, many quantities measured at microwaves are relative, and it is not necessary to know their absolute values. For example, impedance is measured in terms of the characteristic impedance of the transmission line used in the measurement. To ascertain the amount of power reflected or to design a matching network, it is sufficient to know the normalized impedance or admittance rather than the exact number of ohms or mhos. In fact, as will be shown later in Chap. 4, the characteristic impedance of a waveguide can be defined in three different ways, each yielding a different result. Since only the normalized value of an unknown impedance is required, it makes no difference which definition is used.

For power measurements, it is usually sufficient to know the ratio of two powers (or their difference in decibels) and rarely necessary to know power output or input exactly. Of course this is true at low frequencies also; but, at low frequencies, values of absolute power are found so easily that it would be more difficult not to use them. At microwaves, it is much simpler to determine that the power input or output is some unknown constant times the true power. It is not necessary to determine this constant since it cancels out of all relative power measurements.

Besides impedance and power, microwave measurements of phase, frequency, and time are also different from their low-frequency counterparts. However, the absolute values for these quantities are just as important at microwaves as at radio frequencies.

3.1. DETECTORS

Before power can be measured, it is necessary to have some means of detecting its presence. Microwave detectors operate on many different principles; the most common types are *crystals* and *bolometers*.

A microwave crystal is a nonlinear, nonreciprocal device which rectifies the received signal and produces a current proportional to the power input. Since power is proportional to the square of the voltage, the crystal is referred to as a *square-law detector;* its "forward response" is said to obey the square law. It should be noted, however, that this square-law characteristic is only true at low-incident power levels. At powers much higher than ten milliwatts, the characteristic changes gradually to a linear response. In the reverse direction, the crystal presents a high impedance.

The d-c current produced by the crystal can be measured with an ordinary microammeter after the microwave signal has been coupled to the crystal. To do this, the crystal is placed in a mount which can be in a waveguide or coaxial line. The mount usually has some sort of matching elements in it so that the microwave power is absorbed by the crystal and is not reflected. A by-passing capacitance for r-f is also used so that the microwave signal is confined to the crystal mount and is not coupled with the leads connecting the crystal mount to the meter. Finally, it is necessary to provide a d-c return path so that direct current can flow in the circuit. This may be done, for example, by using loop coupling or a shorted matching stub.

The sensitivity of the crystal can be increased by using a modulated signal source, where the signal from the generator is square-wave modulated at an audio rate (usually 1000 cycles) and the crystal output is connected to an audio amplifier which is peaked at the modulation frequency. The output of the amplifier is indicated on a meter which is usually built into the same package.

Bolometers are devices which change resistance when they absorb power. The change can be positive or negative. One type of bolometer contains a short piece of thin metal wire. This type has a positive temperature coefficient of resistivity which means that when it absorbs power its temperature increases and its resistance also increases. It is called a *barretter*. In early microwave systems, ordinary instrument fuses were used as barretters since they contained a thin piece of wire which acted in this manner.

Another type of bolometer is called a *thermistor*. It consists of a small blob of semiconductor material between the input and output wires. The blob is usually a bead, but other shapes such as rods and disks have also been used. Thermistors have a negative temperature coefficient of resistivity so that the resistance decreases as power is absorbed.

Both types of bolometers are square-law detectors. It should be noted that bolometers do not generate currents; thus their mounts must be supplied with a d-c voltage bias across the bolometer. The current is then a function of the resistance which in turn depends on the power. Bolometers are usually used with modulated signal generators and audio amplifiers just as those used for sensitive crystal measurements. The commercial amplifiers used for this purpose have built-in bias sources which can be switched into the circuit. They also have provisions for switching between positive and negative temperature coefficients.

Both thermistors and barretters must be carefully matched in their mounts just as crystals are, and, in general, the same precautions apply to all three. As seen before with crystals, d-c return paths are necessary.

Both types of bolometers are square-law detectors over a wider dynamic range than crystals. Barretters are more sensitive than thermistors, but the latter will withstand severe overloads without deterioration. When using modulated signal sources with bolometers, the modulation frequency is usually about 1000 cycles. This allows a reasonable margin of safety since barretters are useable down to 200 cycles and thermistors down to 100. If the frequency is too low, the heating time-constant allows the resistance to follow the modulation.

A third type of microwave sensing device is a thermocouple. This is an element which generates a d-c voltage when it absorbs microwave power. As with other detecting devices, it must be well-matched, and the mount must have provision for a d-c return. Thermocouples have low sensitivity compared to bolometers and crystals and, consequently, are seldom used as detectors. However, they are usually more reliable than the others for absolute power measurements.

3.2. HIGH-POWER MEASUREMENTS

High-power measurements are always absolute power measurements. They are usually necessary to determine the power output of a high-power tube or of a radar system. As far as measurements are concerned, there is no sharp demarcation between low and high power. Equipment which will measure hundreds of watts is high-power measuring equipment whereas instruments to measure microwatts are obviously low-power units. If the high-power equipment will measure milliwatts and the low-power equipment can be useful up to one watt, there may be considerable overlap.

High-power microwave energy is usually measured by a calorimeter-wattmeter. The power is directed into a well-matched liquid load and is completely absorbed by the liquid. The fluid may be water, oil, or in fact any liquid which is a good absorber of microwaves. The fluid flows through the load and is heated by the microwave energy as it (the fluid) passes

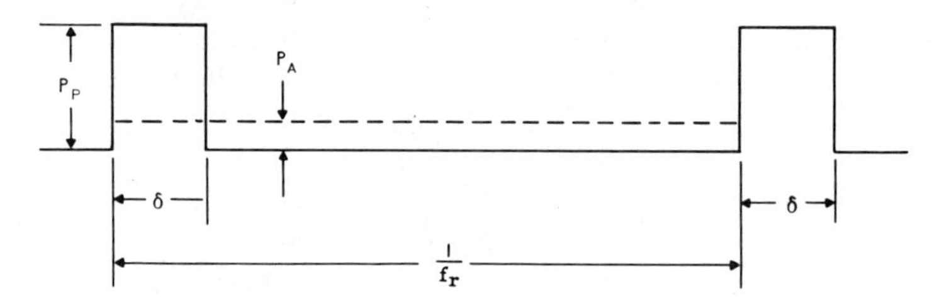

Fig. 3.1. Pulse envelope.

through. The difference between the temperature of the known quantity of liquid before entering and the temperature after it emerges is a measure of the power that has been absorbed. If the rate of flow of the liquid is known, the exact value of the power may be determined. Water loads may be connected directly to a convenient tap, and a flow-meter in the circuit is then used to indicate the flow rate. Oil loads are usually provided with a closed circuit and a pump which pushes the oil through at a predetermined rate. Calorimetric power meters of this type are usually supplied with a meter calibrated directly in power.

The power as measured by a calorimetric power meter is the heating power or *average* power entering the load. In continuous-wave systems, the average power is the only value of power that is generally specified. However, in most microwave systems, the radio-frequency signal is modulated by a pulse. The modulation pulse is nearly rectangular and, in effect, turns on the radio-frequency energy in short bursts. Figure 3.1 shows a typical pulse envelope. Each pulse of energy is transmitted for a time δ which is called the *pulse width*. Pulses are repeated at a frequency f_r which is known as the *pulse repetition frequency* (prf). The time between pulses is then the reciprocal of the pulse repetition frequency or $1/f_r$. The *peak power* is the amplitude of the pulse and is designated by P_p. If the area under a single pulse is spread over the interpulse space, the amplitude would be reduced to a value P_A, which is designated the *average power*. This average power is the amount of continuous-wave power which will produce the same amount of heat as the short pulse. From Fig. 3.1, the area of the pulse is $P_p\delta$. The equivalent area for average power is $P_A(1/f_r)$. Thus,

$$P_p\delta = P_A\frac{1}{f_r}$$

or,

$$P_A = P_p\delta f_r \tag{3.1}$$

Equation (3.1) indicates that the average power is equal to the peak power multiplied by the pulse width and the repetition frequency. The product δf_r is called the *duty cycle*. Thus, if the pulse width is one microsecond (10^{-6} second) and the pulse repetition frequency is 800 times a second, the duty cycle is

$$800 \times 10^{-6} = 0.0008 = 0.08 \text{ per cent}$$

If the peak power with such a duty cycle is one megawatt (10^6 watts), the average power is 800 watts.

The power determined by the calorimeter-wattmeter is average power but it can be converted to peak power by the relationship

$$P_p = \frac{P_A}{\text{duty cycle}} \tag{3.2}$$

assuming that the pulse shape is rectangular. If the pulse is not rectangular its width is that of an equivalent rectangular pulse. Since this is usually evaluated by eye, in practice there may be considerable variation in the values of peak power determined by different individuals.

Calorimetric power meters have been used for measuring powers, from the highest values attainable down to approximately ten milliwatts.

3.3. LOW-POWER MEASUREMENTS

Absolute power below ten milliwatts is usually measured with a bridge circuit in which the resistance of one of the bridge arms is simply a bolometer or thermocouple. Power incident on this arm will change its resistance (in the bolometer) or generate a voltage (in the thermocouple) which will upset the bridge balance. The unbalance of the bridge is then read on a voltmeter which is usually calibrated in terms of power input. For increased accuracy it is possible to introduce audio power to obtain the same unbalance as was obtained with the microwave signal. The microwave power is then equal to this value of audio power, which is easily measured. This eliminates calibration inaccuracy.

This low-power method of measurement may also be used with a suitable coupler to measure higher powers. If a known sample of the incident power is measured with a bolometer bridge, the total power may be calculated by multiplying by the sampling ratio. For example, a 20-decibel coupler is used in the main line from a transmitter. The power out of the coupler is 20 decibels down from the incident power, or one per cent of the incident power. If the reading on the bolometer bridge connected to the coupler is seven milliwatts, the total power must be 100 times this, or 700 milliwatts.

3.4. ATTENUATION

The *attenuation* of a network, component, or device is the ratio of power into it to power out. This is generally expressed in decibels. It should be noted that the input power in attenuation measurements is the

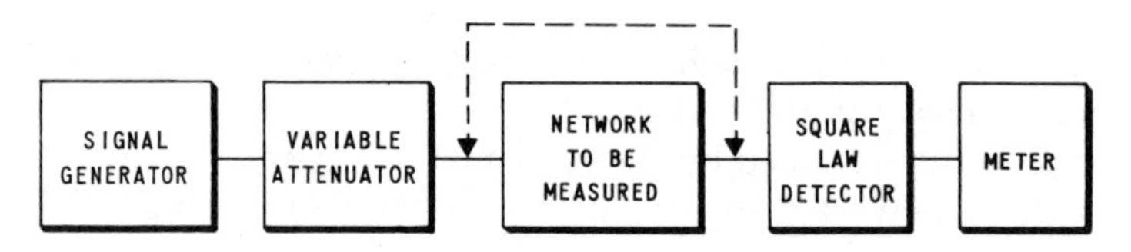

Fig. 3.2. Insertion loss measurement.

power which actually enters the network. If there is a mismatch, there will be some power reflected, but this is not attenuation. The total of power reflected and power attenuated is called *insertion loss*. If there are no mismatch losses, the insertion loss is equal to the attenuation. Since, in practical microwave systems, the VSWR is usually kept low, the losses due to reflection are normally small. If the attenuation is large, there is not much difference between attenuation and insertion loss, and the two terms are used interchangeably. To be entirely correct, however, insertion loss of a network is the difference in power arriving at a load with and a load without the network in the circuit.

Measurements of insertion loss and attenuations are measurements of relative power. Obviously, the insertion loss of a component can be calculated from absolute power measurements with and without the component in the circuit, but this is not necessary. It is simpler to measure the power ratio without worrying about the accuracy of the individual measurements. As long as the detector is square-law, the power ratio can be determined without knowing what the absolute power is. (A typical set-up is shown in Fig. 3.2.) The meter reading is usually set at a maximum with the network out of the circuit by adjusting the variable attenuator in the line. This attenuator may be a separate adjustable pad or it may be built into the signal generator; then the network is put into the line and the reduction in power is noted on the meter. If the meter is calibrated in decibels, the insertion loss may be read directly. If not, it can be calculated from the relationship

$$\text{db} = 10 \log_{10} \frac{P_1}{P_2} \tag{3.3}$$

It must be remembered that with a square-law detector, the *voltage* reading is an indication of *power*.

If a precision calibrated attenuator is at hand, a more accurate method of measurement is shown in Fig. 3.3; the network under test is replaced

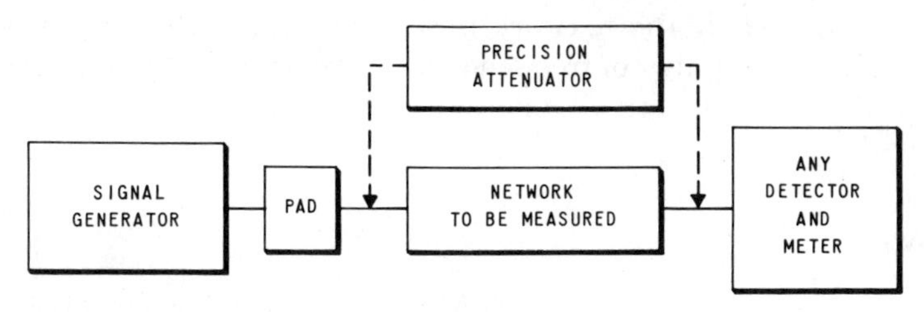

Fig. 3.3. Insertion loss measurement by substitution.

by the amount of attenuation that is necessary to produce the same meter reading. First a reading is noted on the output meter with the network in the circuit. Then the precision attenuator is substituted for the network and adjusted until the meter reading is the same as it was with the network in the circuit. The dial of the precision attenuator now indicates the insertion loss of the network. This method depends only on the accuracy of the attenuator. Since the detector's output reading is the same in both cases, errors in square law or nonlinearity of the meter are of no consequence.

In all power measurements it is necessary to prevent multiple reflections in the line which will cause erroneous readings. For example, in Fig. 3.2, if the output side of the network is mismatched to the line, and the input to the detector is also mismatched, part of the signal will bounce back and forth between the two discontinuities. The relative phases of the reflected signals will depend on their spacing in wavelengths and, thus, will vary with frequency, adding at some frequencies and canceling at others. If the detector were well-matched to the line this would not happen. In practice, however, it is not possible to match a detector to the line over a broad band of frequencies. Reflections can be reduced, though, by placing a matched resistive pad in the line between the network and the detector. This will absorb the reflections and eliminate this source of error. A six-db pad is usually sufficient.

Obviously, the frequency and power output of the signal generator must remain constant during the course of the measurements or the readings will be meaningless. To insure control, a directional coupler may be placed in the circuit as close to the signal generator as possible. A detector on the output of the coupler should remain constant during the measurements. If not, it warns of an error, and measurements must be repeated.

In making attenuation measurements, the signal generator or signal source is set to the desired frequency. If readings are required for a band of frequencies, measurements are repeated at suitable intervals in this band. Normally, measurements are made at the band edges and center frequency; however, if the network is so complicated that its response may be erratic, the number of frequency measurements must be increased. This procedure is true for all microwave measurements that may be frequency-dependent.

3.5. VSWR

Standing waves are an indication of the quality of transmission. A well-matched line has no reflections and consequently the VSWR is unity. Measurement of VSWR is the most frequent measurement at microwaves and one of the most important.

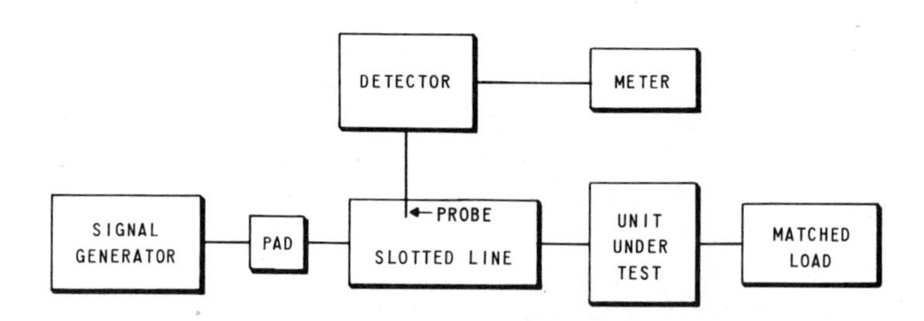

Fig. 3.4. Standing-wave measurement.

Figure 3.4 illustrates a typical VSWR measuring set-up. The slotted line is its basic instrument. A probe is moved along the line to sample the voltage; the output of the probe is detected and read on a meter. The ratio between maximum voltage and minimum voltage is, of course, the VSWR.

The signal source must present a matched impedance looking back into it. Otherwise, reflections from the network being tested will reflect again off the signal generator and cause the peaks and nodes to shift position in the standing-wave pattern. To match the generator a pad is put between it and the rest of the equipment. The pad can be resistive and absorb some of the reflections; but it also attenuates the input signal. If the signal source is at too low a power level to permit additional attenuation, a resistive pad cannot be used because the minimum readings will be lost in the noise of the instruments. To prevent that loss, a ferrite isolator is used as a pad. This is a component which permits signals to pass in one direction with low attenuation, but absorbs power in the reverse direction. It is an ideal pad for VSWR measurements as well as for other laboratory measurements.

The standing-wave ratio as measured will be the VSWR at the input of the component under test. If the component does not absorb all the power that enters, it must be terminated in a matched load as indicated in Fig. 3.4. Otherwise, the output would present a mismatch to free space and the subsequent reflection would be detected at the slotted-line probe along with the desired reflection. If the component being tested is an absorbing device, such as a load, then a matched load on the output is unnecessary.

The signal source must put out a single frequency. If the signal is frequency-modulated or if there are spurious signals present, each frequency will produce a different VSWR and a different maximum and minimum position. This will lead to erroneous readings. Sine-wave modulation of a klystron in the signal source can produce a frequency-modulated signal. For proper measurements, the source should be square-wave-modulated.

The probe entering the slotted line is in itself a discontinuity. In order to hold its disturbing effect to a minimum, the depth of penetration should be kept as small as possible. At low standing-wave ratios, this is no problem, but for higher values the probe must be inserted deeper so that the minimum can be read above the noise level. The disturbing effect of the probe is such that as it moves deeper it introduces an error which produces a value of VSWR lower than that which exists. This error becomes appreciable for values of VSWR above ten, but below ten the VSWR may be read with reasonable accuracy by holding the probe depth to a minimum.

For high values of VSWR, the *twice-minimum* method should be used. In this method the probe depth does not introduce an error; the probe is inserted to a depth where the minimum can be read without difficulty, and then moved to a point where the *power* is twice the minimum. This position is denoted d_1. The probe is moved to the twice-power point on the other side of the minimum—the position designated d_2. The VSWR may be found by the relationship

$$\text{VSWR} = \frac{\lambda_g}{\pi(d_1 - d_2)} \tag{3.4}$$

The wavelength and distance must be in the same units.

In Eq. (3.4), λ_g designates the wavelength in the slotted line. For an air-filled coaxial line, λ_g is the same as the free-space wavelength. In a waveguide, however, λ_g is the guide wavelength which is different from that of free space. This will be explained in detail in the chapter on waveguides; for the present, λ_g can be found in tables available in the *Microwave Engineers' Handbook* and elsewhere.

It should be noted that the probe is moved to the twice-*power* points. If a standing-wave indicator is used, it is probably calibrated for use with a square-law detector, and consequently the ratio of the two readings would be 1.4:1 or three decibels.

The output meter at the detector in Fig. 3.4 can be any kind of voltmeter; in practice, however, a standing-wave indicator is used. This is an audio amplifier peaked at the modulation frequency with a meter calibrated especially for reading VSWR. The probe is set at a voltage maximum, and pads are adjusted so that the meter reads full scale. The probe is then moved to a minimum point, and, assuming there is a square-law detector, the scale on the meter reads VSWR directly.

If the detector is replaced, the new one should be checked for square-law response by leaving the probe at one position, increasing the power level in known increments, and observing the output on the standing-wave indicator.

3.6. REFLECTOMETER

VSWR may be measured without using a slotted line by using two directional couplers instead. (Directional couplers will be considered later in detail in Sec 7.6.) For the purpose of making VSWR measurements, a directional coupler is a component which samples the energy going in one direction in the main line without this being disturbed by energy going in the reverse direction. Figure 3.5 shows a *reflectometer* set-up. The directional couplers need not have identical coupling characteristics. If the input coupler has 40 decibel coupling, the energy at detector No. 1 will be 40 db down or one ten-thousandth of the power out of the signal generator. Similarly, if the output coupler has 20 decibel coupling, the energy at detector No. 2 will be one-hundredth of the reflected power. Due to the directional properties of the couplers there is no interaction between the two readings. Comparison of the outputs at the two detectors gives the reflection coefficient. From the reflection coefficient, the VSWR may be determined by using Eq. (2.24).

It is conceivable that the two outputs could be combined in some sort of circuitry which would read their ratio directly. Such an instrument

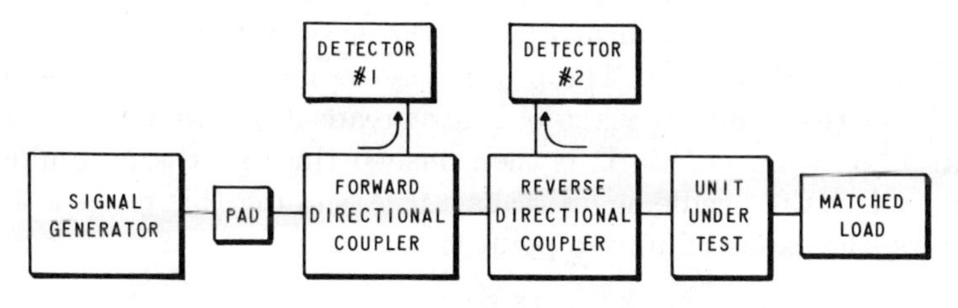

Fig. 3.5. Reflectometer.

does exist. It is called a *ratiometer* and its output meter is calibrated directly in VSWR.

3.7. IMPEDANCE

In Chap. 2, Eq. (2.40) demonstrated that the value of an unknown impedance could be calculated from the VSWR and the distance of a minimum position from the unknown impedance.

$$z_s = \frac{z_L + j \tan \beta s}{1 + jz_L \tan \beta s}$$
$$y_s = \frac{y_L + j \tan \beta s}{1 + jy_L \tan \beta s} \tag{2.40}$$

At the minimum position the voltage and current are both real, and, consequently, so is the normalized impedance z_s. At a minimum, z_s is the reciprocal of the VSWR. (At a maximum, z_s equals the VSWR.) The quantity βs is simply the electrical distance (length in wavelengths) between the position of the minimum and the unknown impedance. The equation can now be solved for the unknown load impedance z_L. In general, z_L is complex. In practice, however, all the calculations can be avoided by using a Smith Chart.

The set-up for making impedance measurements is the same, illustrated in Fig. 3.4, as for VSWR measurements. The unknown impedance is attached to one end of the slotted line, and the VSWR is carefully measured. For impedance it is necessary to note the position of the minimum *as well as* its amplitude. For high standing-wave ratios, the minimum position is usually well defined, but for low ratios, the minimum is broad and thus difficult to locate exactly. A useful method to circumvent this difficulty is to determine the positions of two equal voltage points on either side of the minimum, and to assume the minimum is half-way between them. These points should be closer to the minimum than they are to the maximum. Thus, assume that the VSWR is two to one and that the minimum is somewhere near the 15-centimeter mark on the slotted line scale, but is too broad to read accurately because it is changing too slowly at this point. The probe may be moved to one side until the needle on the standing wave indicator reads 1.9. Assume the probe is now at 14.62 centimeters. It is then pushed through the minimum to the point where the needle again reads 1.9. Assuming this point is 15.52 centimeters, the exact minimum point is

$$\frac{14.62 + 15.52}{2} = 15.07$$

A short circuit is now attached to the slotted line in place of the load being measured, and a new minimum is found. This is called the *reference minimum* and the position of the short is called the *reference plane*. When the value of impedance is finally determined by using the reference minimum, it will be the impedance at the reference plane.

Continuing the example of the preceding paragraph, assume the reference minimum occurs at 16.08 centimeters and the frequency is 2800 megacycles. At this frequency, the wavelength, λ, is 10.72 centimeters. When a short circuit is placed at the end of a line, there is a voltage minimum there and at half-wavelength intervals from it. It may be noted that 16.08 is three half-wavelengths at this frequency. The difference in position between the reference minimum and the impedance minimum is called the *shift* and is expressed in wavelengths. Thus the shift is

$$16.08 - 15.07 = 1.01 \text{ cm} = \frac{1.01}{10.72} = 0.094\,\lambda$$

The *direction* of the shift is the direction from the impedance minimum to the reference minimum and is expressed as being toward the load or toward the generator. In the present example, with the zero-end of the slotted line toward the load, the shift was from 15.07 to 16.08, or toward the generator. Now the impedance can be plotted on the Smith Chart, as shown in Fig. 3.6.

Since the shift is toward the generator, the rotation is clockwise on the chart, as indicated by the arrow at the left. A line is drawn from the center to the point on the circumference indicating a shift of 0.094λ toward the generator. To open a compass to a VSWR of two to one, its fixed leg is put at the center and its writing leg at 2.0 on the diameter. This point, 2.0, is the maximum impedance. It's reciprocal is the impedance a quarter-wavelength away which is the minimum and diametrically opposite.

The arc on Fig. 3.6 represents the rotation from the minimum to point P, the unknown impedance. Point P shows that the unknown

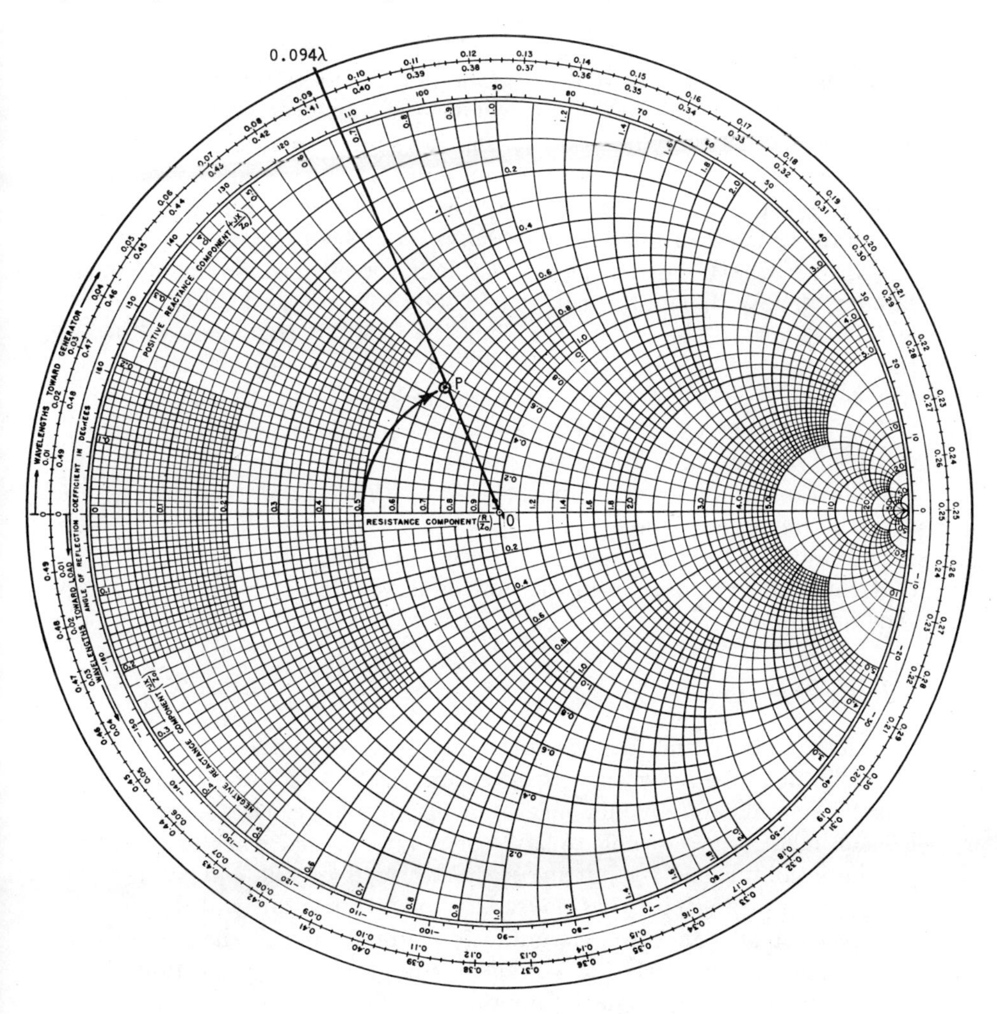

Fig. 3.6. Impedance plot.

impedance is $0.65 + j0.45$. This is, of course, normalized and must be multipled by the impedance of the slotted line to get the unknown impedance in ohms.

Obviously, it makes no difference whether the reference minimum is taken before or after the unknown impedance. It is only necessary to make sure that both are at the same frequency. By keeping the shift small, the accuracy of the frequency reading is not too important. For example, if the frequency were off two per cent, the shift might be 0.092λ instead of 0.094λ. Referring to the Smith Chart it can be seen that this would introduce a negligible error in the impedance reading.

If the zero on the slotted-line scale is right at the flange and if the frequency is known accurately, it is not necessary to determine the reference minimum. It may be assumed that if a short circuit is placed at zero, zero may be taken as the reference minimum. In the example given, 15.07 is the impedance minimum. Since zero is toward the load from this point, the shift is 15.07 centimeters or 1.406λ ($= 15.07/10.72$) toward the load. The Smith Chart repeats itself every half wavelength, so 1.406λ is the same as 0.406λ. It can be seen that 0.406λ toward the load is the same as 0.094λ toward the generator. Thus the result is the same. However, the accuracy of the frequency reading is more important with higher wavelengths. A two per cent error in frequency could make the shift 1.430λ instead of 1.406λ, which results in a sizable error in impedance.

It should be noted that, using the reference minimum method, it is unimportant whether zero on the slotted-line scale actually coincides with the reference plane or, in fact, whether the slotted line has the high or low numbers toward the load. The important thing to remember is that the shift is always from the impedance minimum toward the reference minimum.

3.8. FREQUENCY

The slotted line is a useful component for measuring frequency since the distance between voltage minima is known to be a half-wavelength. To make a frequency measurement, or more exactly a wavelength measurement, it is desirable to have the minima as sharp as possible. This can be accomplished by putting a short circuit on the end of the line, although anything else that causes complete reflection will do as well. The probe may penetrate the guide to a greater depth than in VSWR measurements since relative voltage readings are not important. It is customary to read as many minimum voltage positions as possible and average the spacing between them. Twice this average spacing is equal to a wavelength. The frequency may be found then from the relationship $f = c/\lambda$ where c is the speed of light if the measurement is in coaxial line, or from suitable tables if the measurement is in waveguide.

A better way to measure frequency is with a calibrated resonant cavity. A resonant cavity-wavemeter is the microwave analog of a tuned resonant circuit. In general there are two primary types: (1) transmission cavities, which pass only the signal to which they are tuned, and (2) absorption cavities, which attenuate (by absorption) only the frequency to which they are tuned. The absorption type is preferred for laboratory frequency measurements. A typical set-up is shown in Fig. 3.7. First the power level is adjusted to give a full-scale reading on the output meter; then the wavemeter is tuned slowly until there is a dip in the power level. The frequency may then be read from the dial of the wavemeter.

Of course, the slotted line in Fig. 3.7 is not necessary to measure frequency, but it doesn't interfere with the frequency measurement. Since reading frequency is often necessary when making VSWR measurements, Fig. 3.7 shows a set-up that will do both. In fact, if the pad is replaced

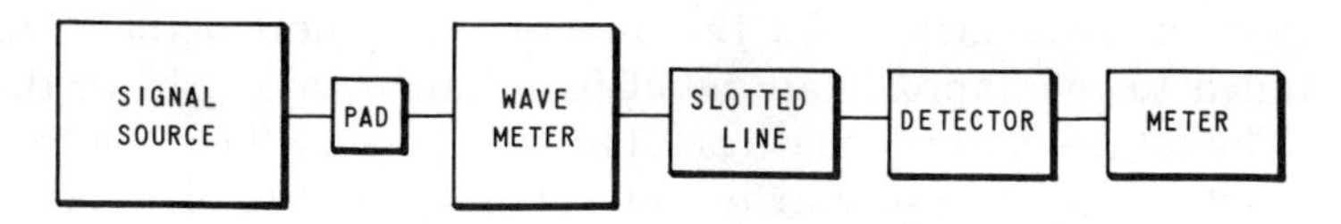

Fig. 3.7. Frequency measurement.

with a calibrated attenuator, the set-up may be used for insertion loss measurements as well.

In all the measurements described thus far, a signal generator has been shown as the signal source. It has a dial calibrated in frequency which is accurate enough for most laboratory measurements. When more accurate frequency readings are demanded, a wavemeter is also used. If no signal generator is available, a klystron (with its own power supply) is used as a source of microwave energy in the laboratory. With a klystron, a wavemeter is required to read frequency simply, although for lack of alternatives the slotted line could be used.

Each measurement, whether of attenuation, impedance, or anything else, is made at a particular frequency. If it is required to have a plot of attenuation as a function of frequency, or VSWR as a function of frequency, it is necessary to make a series of individual measurements, noting the frequency at which each measurement is made.

3.9. PHASE

It is sometimes necessary to know the phase shift through a network. This is equivalent to knowing the electrical length of the network; that is, its length in wavelengths. In all methods of measuring phase, it is assumed that the worker has an approximate idea of the length, since it is impossible to distinguish between one quarter-wavelength and five quarter-wavelengths as far as phase is concerned.

A simple measurement of phase uses a slotted line and a short circuit. The short is placed at the end of the line and a reference minimum is noted. Then the unknown network is attached to the slotted line, and the short is placed at the output of the network. A new minimum will be noted. Since this minimum is a multiple of half-wavelengths from the short, it should be possible to calculate the exact electrical length of the network. For example, suppose at 2800 megacycles a reference minimum is found at 16.08 centimeters. (At 2800 megacycles, λ is 10.72 centimeters.) Now suppose that a network approximately two wavelengths long is placed at the end of the line, and with a short at the end, the new minimum is at 14.90 centimeters. The short has apparently moved 1.18 (= 16.08 − 14.90) centimeters. This is $0.110\lambda(1.18/10.72)$. Since this apparent movement is in the direction the short actually moved, it is added to the approximate number of half-wavelengths in the network. The total electrical length is 2.110 wavelengths. This can be multiplied by 2π to give phase shift in radians or by 360° to give phase shift in degrees. Thus,

$$2.110\lambda = 4.22\pi \text{ radians} = 759.6°$$

Sometimes it is required to note the change in phase (again, this means change in electrical length) of a component or network as some parameter is varied. For example, the phase changes as a piece of dielectric material is moved from the edge of a waveguide toward the middle. It might be useful to know the change in phase as a function of position. In this case the component is attached to the slotted line, as in the last example, and is terminated in a short circuit. The dielectric piece is positioned against one wall of the waveguide, and the position of the minimum is noted. The dielectric is now moved a small distance toward the center. This in effect lengthens the component so that the short circuit is farther away. The minimum position will follow since it must stay an integral number of half-wavelengths away from the short circuit. The electrical distance that the minimum moves in the guide is the change in electrical length for that movement of the dielectric. This can be repeated and plotted to obtain a graph of phase shift as a function of position. Since this is done in small steps, it is possible to calibrate more than 360° of phase shift in this way, even though the minimum positions repeat every half-wavelength.

Some components are unilateral or nonreciprocal. If it is necessary to measure the phase shift through a ferrite phase shifter, for example, the short-circuit method is inadequate, since it determines the total phase shift through the unit in two directions. (The signal passes through the unit, is reflected from the short circuit, and passes back through the unit.) In a ferrite phase shifter, the phase shift in one direction is different from what it is in the reverse direction. Therefore, in order to determine each of these phase shifts separately, a different approach is necessary. The set-up for measuring one-way phase shift is shown in Fig. 3.8. The power is fed into a power splitter which divides the signal into two approximately equal signals. Pads are placed in each arm to make sure they are properly matched and to prevent interaction. Nonreciprocal pads, such as ferrite isolators, are preferable, but ordinary resistive attenuators are satisfactory.

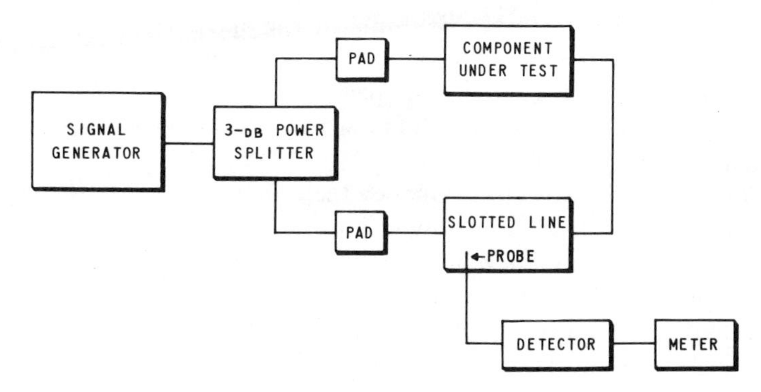

Fig. 3.8. Phase shift measurement.

In the slotted line, two equal or nearly equal signals are traveling in opposite directions. Since they are at the same frequency, there is no way to distinguish then from an individual signal undergoing total reflections. Thus, there will be deep nulls at half-wavelength intervals. A typical problem might be to determine the variation in phase shift in a ferrite component as a function of current through its coil. First a minimum is noted with zero current. Then as the current is varied in small steps, new minimum positions are recorded. If the component under test becomes a half-wavelength longer, the minimum will move only a quarter-wavelength toward the component. The other path to the probe is lengthened a quarter-wavelength while the path through the component is shortened this much. The net change is a half-wavelength. Similar reasoning indicates that the phase shift using this set-up is twice the shift in the null position. In other words the shift in centimeters is divided by a half-wavelength instead of by a full wavelength.

An alternative method is to place a calibrated phase shifter at point X in Fig. 3.8. Then as the phase through the test piece is varied, the calibrated unit is changed with the minimum remaining unchanged. The required phase shift can then be read directly from the calibrated phase shifter.

3.10. SWEEP OSCILLATORS

All of the measurements described thus far have been indicated as single frequency checks. If it is required to determine VSWR, or phase, or attenuation, or anything else, as a function of frequency, it is necessary to change the frequency of the signal source and plot new points at each frequency setting. This is called a point-to-point method of measurement and is necessary whenever accurate, quantitative measurements are required. For rapid evaluation of a component, however, qualitative results are satisfactory. For example, if the specification states that the VSWR must be under 1.25, it is not necessary to know whether it is 1.10 or 1.20 as long as it meets specifications. In such cases swept frequency measurements are much faster than point-to-point measurements and are equally reliable.

The heart of the swept frequency technique is a sweep oscillator which changes frequency electronically at a rapid rate. Sweep oscillators usually cover an octave frequency band; that is, one to two kilomegacycles, two to four, etc. They have provisions to sweep any smaller portion of the octave band if the whole octave is not of interest. The sweep rate is usually variable from one sweep in more than two minutes up to fifty or more sweeps in a second.

Since it is not feasible to tune each element in the circuit to each frequency as the oscillator is swept, all components used must have broad-band characteristics. The final meter, however, must be replaced with something which will display the result as a function of frequency. This is usually an oscilloscope or a recorder. In both cases, the horizontal motion is synchronized with the saw-tooth voltage which sweeps the oscillator. The signal output is detected in the usual way, fed first into an amplifier, and afterwards to the vertical motion of the oscilloscope or recorder. The trace will then be a plot of signal as a function of frequency.

The broad-band components used with the sweeper are never flat with frequency. For example, the response of a detector is usually optimized at two or three frequencies in the band in order to simulate a multiple resonant tuning. In addition, the swept oscillator itself maintains a minimum power output, but this output varies with frequency. Consequently, it is necessary to calibrate the face of the oscilloscope with a known mismatch or known attenuation.

Figure 3.9 shows a set-up for making VSWR measurements with a sweep oscillator. The directional coupler is reversed so that only reflected signals arrive at the detector. Thus, the oscilloscope presentation will show reflection as a function of frequency. Before making the test it is necessary to calibrate the oscilloscope. This can be done by replacing the test piece and its load in Fig. 3.9 with standard reflections. For example, it might be replaced with something that has a known VSWR of 1.1:1 over the frequency band. The trace on the oscilloscope can be marked with a grease pencil. Similar traces are drawn for mismatches that bracket the expected mismatch of the test component and the specification VSWR.

Another method avoids using standard mismatches, which are usually difficult to obtain. Instead, a calibrated attenuator is used to reduce the

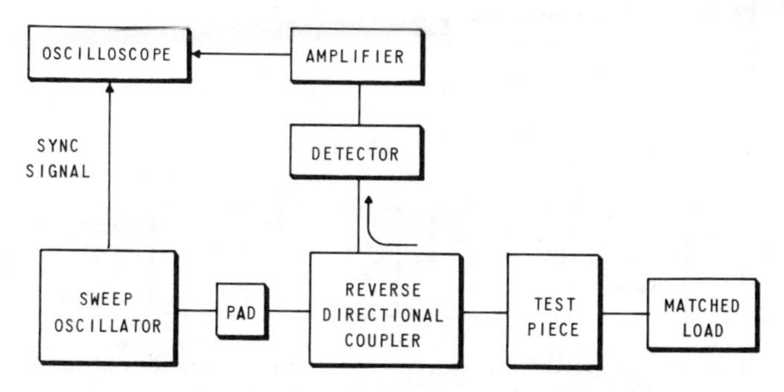

Fig. 3.9. VSWR measurements using sweeper.

reflection from a short circuit to the equivalent reflection from the desired VSWR. The relationship between the amplitude of the voltage reflection coefficient $|\Gamma|$ and the VSWR was shown in Eq. (2.25) to be

$$|\Gamma| = \frac{\text{VSWR} - 1}{\text{VSWR} + 1} \tag{3.5}$$

The quantity $|\Gamma|$ is expressed as a dimensionless number, the ratio of the voltage reflected to the voltage incident. However, the same information could be presented by referring to the loss in decibels between the incident and reflected signals. This quantity is called the *return loss* and is designated L_R. The relationships are

$$L_R = 20 \log_{10} \frac{E_1}{E_R} = 20 \log_{10} \frac{1}{|\Gamma|} = 20 \log_{10} \frac{\text{VSWR} + 1}{\text{VSWR} - 1} \tag{3.6}$$

Thus, for a VSWR of 1.25, $L_R = 20 \log_{10} \frac{2.25}{0.25} = 19.09$ decibels. The set-up for this method is shown in Fig. 3.10. First the shorting switch is put in the short-circuit position so that there is total reflection. (If no shorting switch is available, a simple short circuit can be attached at this point.) The calibrated attenuator is set at values determined from

Eq. (3.6) to reduce this total reflection to desired values of VSWR and traces are marked on the face of the scope. Then the shorting switch is opened and the attenuator set at zero so that the reflection from the test piece reaches the detector. (A typical presentation is shown in

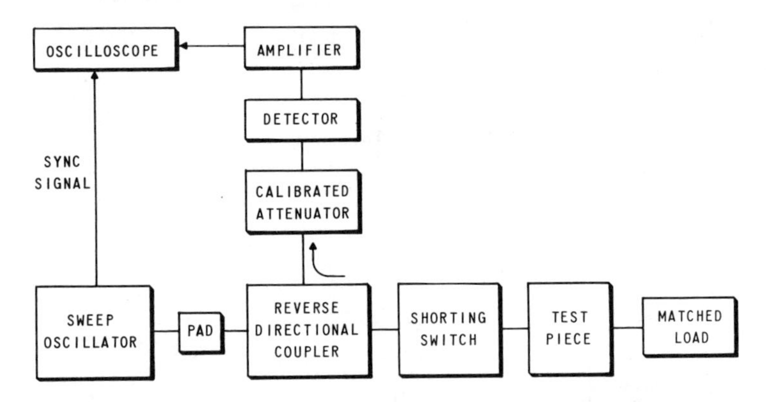

Fig. 3.10. Swept VSWR measurements using calibrated attenuator.

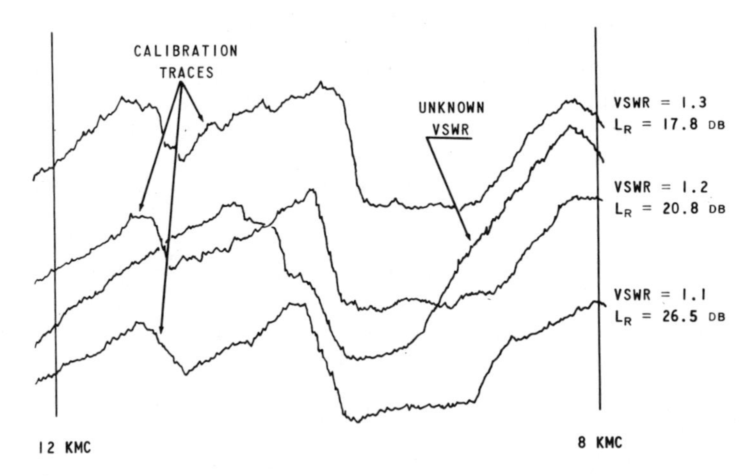

Fig. 3.11. Presentation of VSWR using swept oscillator.

Fig. 3.11.) Calibration curves are shown for standing-wave ratios of 1.1, 1.2, and 1.3. The amplitude variations are caused by variations in the detector, the oscillator output, and the directional coupler. The trace for the unknown VSWR after calibration shows a value near 1.3 at the low frequency end, dropping to about 1.15 at the center and having another peak of about 1.24 near 11 kilomegacycles. If the specification is 1.3, the component would pass. Obviously, the same sort of presentation would appear if a recorder were used instead of an oscilloscope. It should be noted that the calibration does not have to be repeated, but should hold for subsequent measurements as long as the gain controls, etc, remain fixed.

If a reflectometer set-up, as shown in Fig. 3.5 is used with a swept oscillator, the output of the two detectors can be fed into a ratiometer,

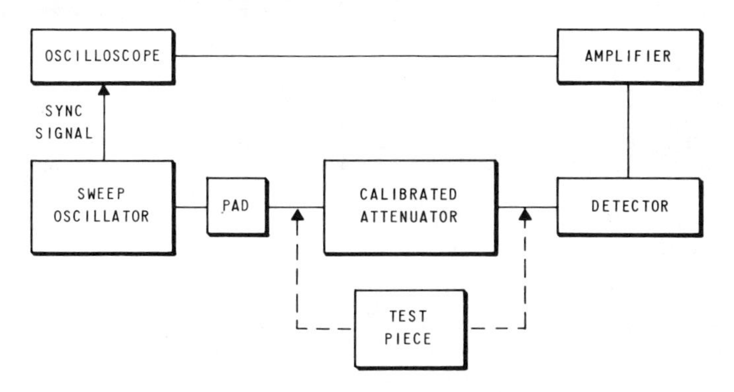

Fig. 3.12. Attenuation measurements using sweeper.

the output of which appears on the synchronized oscilloscope. Since the output of the ratiometer is the ratio of reflected power to incident power, variations in power output of the oscillator will cancel out and the calibration lines of the oscilloscope presentation will thus be almost straight lines.

Attenuation can also be measured with a sweeper. A simple set-up is shown in Fig. 3.12. Using a calibrated attenuator, the circuit traces are made on the oscilloscope with values of attenuation which will bracket the expected attenuation. For example, if the specification calls for a minimum of twenty decibels of attenuation, traces might be made with values of attenuation such as 15, 20, 25, and 30 decibels. A grease pencil can be used to mark these on the face. The calibrated attenuator is removed and the unknown attenuation substituted. The new trace appearing on the face of the oscilloscope can be interpreted, "attenuation as a function of frequency." Strictly speaking, this trace represents insertion loss rather than attenuation, but, as previously mentioned, the difference between the two is negligible, when the attenuation is high. As with VSWR measurements, a reflectometer smooths the lines but is not a necessity.

QUESTIONS AND PROBLEMS

3.1. Discuss the characteristics of two types of bolometers.

3.2. What factor limits the lowest frequency of the modulation envelope when a bolometer is used?

3.3. Describe two methods for measuring high r-f power levels.

3.4. The prf of a radar transmitter is 1000 per sec. The peak pulse power is 1 Mw. The pulse width is 0.5 μsec. What is the average power level? What is the duty cycle?

3.5. Distinguish between attenuation and insertion loss.

3.6. What changes would you make in the setup of Fig. 3.2 if the measurements were to be made over a band of frequencies? Show the proper block diagram.

3.7. Why is a "high quality" square wave necessary when it is used to modulate a klystron signal source for microwave measurements?

3.8. A slotted line is being used to check the VSWR on a 50-ohm air-dielectric line. The frequency is 200 Mc. The VSWR is high so the twice-minimum method is to be used. The distance between points of twice minimum power is found to be 5.3 cm. What is the VSWR on the line? What is the magnitude of the reflection coefficient?

3.9. Explain how you could check the square-law response of a crystal that is to be used in connection with VSWR measurements.

3.10. Two identical directional couplers are placed in a waveguide to sample the incident and reflected power. The meter shows the power level of the reverse coupler to be 10 db down from the level of the forward coupler. What is the VSWR in the guide? *ans.* 1.92

3.11. A coaxial air-dielectric line is being used to measure the value of a load. The VSWR on the line is 3, and the signal frequency is 2 kMc. When the load is replaced with a short the voltage minimum moves 3 cm toward the generator. What is the impedance of the load?

3.12. You have the test set-up as shown below. The short is put in the guide and you read +5 db. The short is now removed and the power meter reading is −9.5 db. What is the magnitude of the reflection coefficient and the VSWR in the guide? *ans.* 0.18 and 1.47

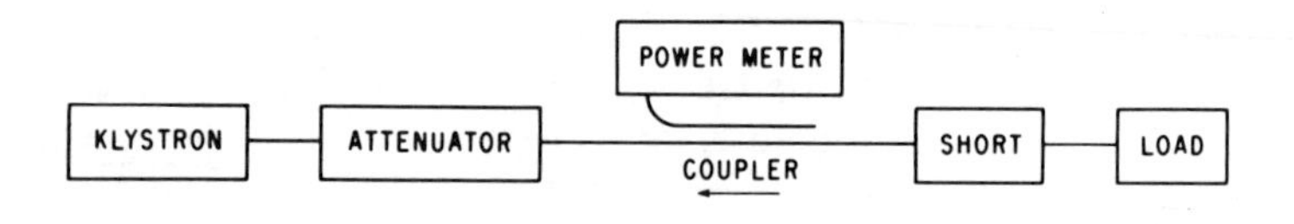

Fig. P 3.12

3.13. A load test is being run with a reflectometer as shown below.
a. What is the return loss? *ans.* a. 17 db
b. What is the magnitude of the reflection coefficient? *ans.* b. 0.14
c. What is the VSWR? *ans.* c. 1.32

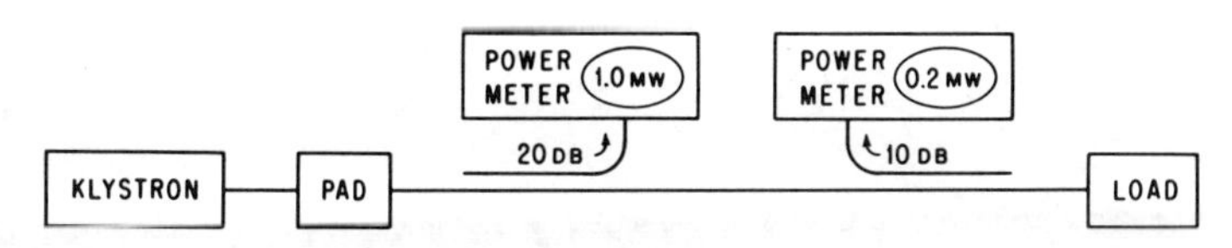

Fig. P 3.13

3.14. The test set-up shown in Fig. 3.8 is being used to check the effect on phase of moving a piece of dielectric in a wave guide. The frequency is 10 Gc. The wavelength in the guide is 4 cm. It is found that the position of a null can be moved 1.1 cm. What is the maximum phase shift in degrees?

4

WAVEGUIDES

To the engineer accustomed to working at low frequencies, a waveguide doesn't look like a transmission line. It seems impossible for electromagnetic waves to travel down a hollow pipe which appears to short-circuit both sides of the electric circuit. Waveguides are not feasible at low frequencies because circuit dimensions are small compared to a wavelength. Waveguides are used when their dimensions are practical fractions of

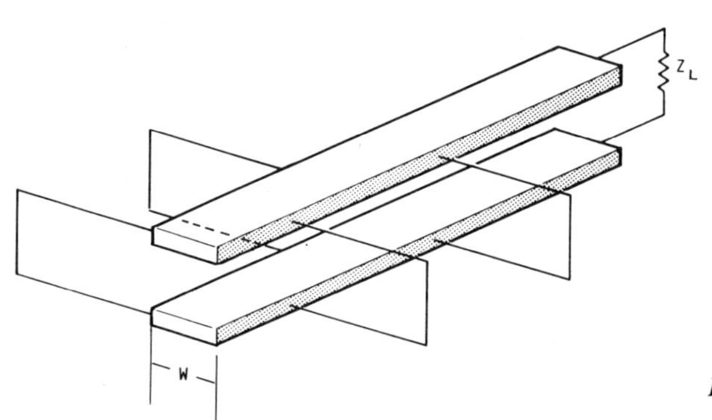

Fig. 4.1. Development of a waveguide.

wavelengths or larger. This accounts for the difference between frequencies, for even at low frequencies, short-circuited quarter-wave stubs are used to appear as open circuits, and there is no difficulty understanding them.

As a matter of fact, a plausible explanation for propagation in a waveguide can be derived from the idea of quarter-wavelength short-circuited stubs. Figure 4.1 shows how such a waveguide might be developed. Two metal bars of width w form a two-wire transmission line to deliver power to a load Z_L. In Fig. 4.1 the two bars are supported by four quarter-wave short-circuited stubs—two on each side. These stubs appear as open circuits across the transmission line. They have no effect on the propagation along the line because an infinite impedance in shunt with a line produces no change. If the number of stubs is doubled, there will still be no change, and, in fact, if an infinite number of such stubs were used so that they touched each other, there would still be no change. With the stubs touching each other, the circuit would have the configuration of a waveguide. If the frequency is raised, it is easy to imagine that the width w is increased so that what remains of the total waveguide width presents a quarter-wavelength on each side of the bars. If the frequency is lowered, w, of course, has to be decreased. However, when the frequency is lowered to the point where the width of the "waveguide" is less than a half-wavelength, even if w were decreased to zero, it would be impossible to have quarter-wavelength stubs on either side of the center. It will be shown later that propagation in this case would be difficult or impossible.

4.1. MODES

Although the explanation above may satisfy the person working at low frequency who is accustomed to currents on wires and voltages between wires, it is too cumbersome for microwave work. At microwaves it is usually simpler to consider magnetic and electric fields. These are always perpendicular to each other, and together form an electromagnetic wave which travels through the waveguide. As the wave moves in the waveguide, it can have an infinite variety of patterns, which are called *modes.*

The fields in the waveguide which make up these mode patterns, and, in fact, all electromagnetic fields, must obey certain physical laws. As was mentioned, the electric field is always perpendicular to the magnetic field. At the surface of a conductor, the electric field cannot have a component parallel to the surface. This indicates that the electric field must always be perpendicular to the surface at a conductor, although it is not necessarily perpendicular a short distance away from the conductor. The magnetic field, on the other hand, is always parallel to the surface of a conductor and cannot have a component perpendicular to it at the surface, although such a component can exist a short distance from the surface.

The simplest mode in rectangular waveguides is shown in Fig. 4.2. The solid lines are voltage lines and indicate the direction of the electric field; the dotted lines, usually called *H lines,* indicate the magnetic field. The fields are indicated at one moment of time. As shown in the end view, there is a voltage between the top and bottom of the guide which is maximum at the center. (This is demonstrated by the placement of more arrows in the center than toward the edges.) It is zero at the side walls, since these walls short out the voltage. A quarter of a cycle later, the voltage is zero, and again a quarter of a cycle later, the voltage will appear as in the figure but with the arrows reversed. This is indicated by the dots (in the top view) representing the points of the arrows, and the x's representing the tails. The magnetic field as shown by the dotted lines consists of closed loops. It should be noted that as the wave propagates along the waveguide, the electric and magnetic fields move together.

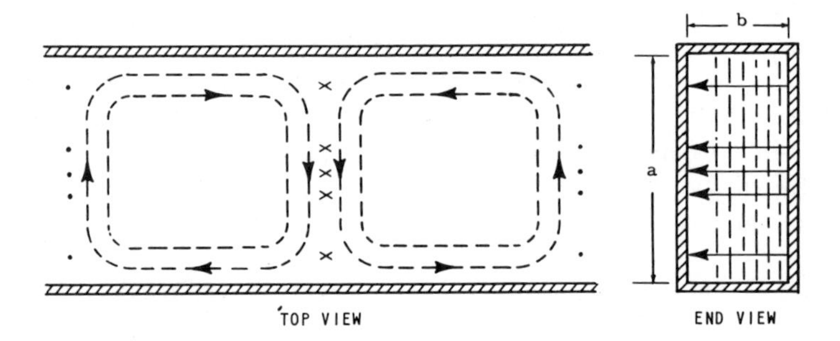

Fig. 4.2. Dominant mode in rectangular waveguide.

In general, there are two kinds of modes in a waveguide. In one type, the electric field is always transverse to the direction of propagation, as it is in Fig. 4.2. This type of mode is called a TE or *transverse electric* mode. In the second type of mode the magnetic field is always transverse to the direction of propagation. This is a TM or *transverse magnetic* mode. Thus, in a TE mode, no electric line is in the direction of propagation; and, of course, the same restriction holds for magnetic lines in TM modes.

Two subscripts are used to designate a particular mode. The first subscript indicates the number of half-wave variations of the electric field (even in a TM mode) across the wide dimension of the waveguide, and the second indicates the number across the narrow dimension. Referring to Fig. 4.2, it can be seen that the voltage varies from zero to a maximum to zero across the wide dimension. This is one half-wave variation. Across the narrow dimension (going from top to bottom) there is no variation in voltage. Hence, this mode is the $TE_{1,0}$ mode. End views and top views of three other modes are shown in Fig. 4.3.

It is not necessary for the waveguide to be rectangular in cross section; in fact, almost any shape will support electromagnetic waves. Irregular shapes would be too difficult to analyze and are rarely used, but round waveguides are quite common.

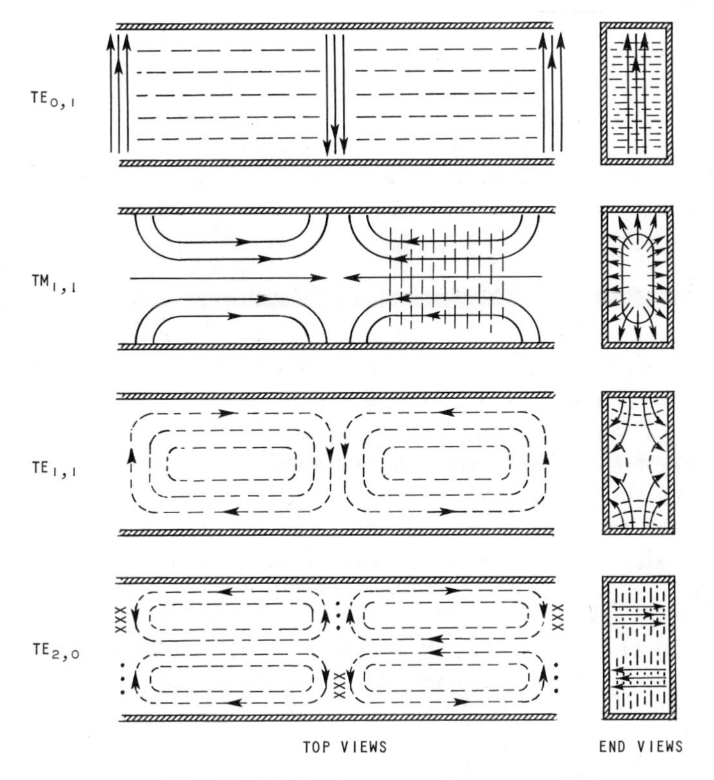

Fig. 4.3. Modes in rectangular waveguide.

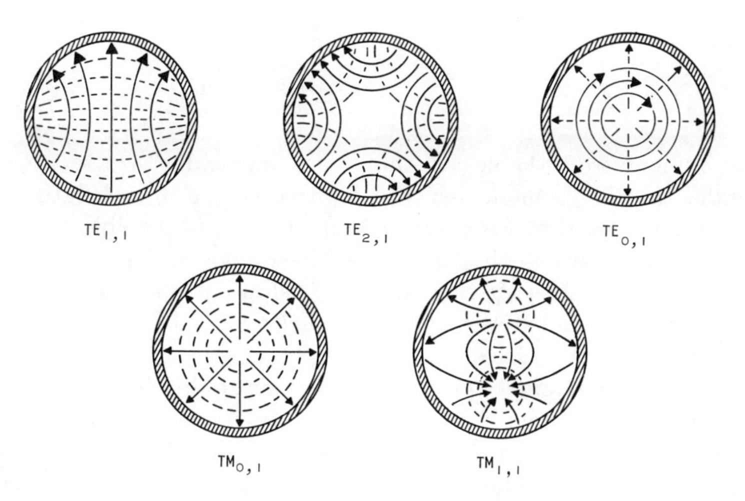

Fig. 4.4. Modes in circular waveguide.

In circular waveguides there are TE and TM modes as in other waveguides; however, the circular waveguide subscripts have different meanings. Because they do not have a simple meaning as in rectangular waveguides, it is easier just to remember what the individual modes look like. Figure 4.4 shows the end views of some common circular waveguide mode patterns.

4.2. CUT-OFF WAVELENGTH

For each mode of operation there is a *cut-off wavelength*, λ_c, determined by the dimensions of the waveguide. For the $TE_{1,0}$ mode in rectangular waveguide, $\lambda_c = 2a$, where a is the wide dimension of the waveguide. (The narrow dimension is denoted b.) This means that at frequencies where the wavelength is longer than twice the wide dimension, there can be no propagation in the waveguide. These frequencies are said to be *below cutoff*. Table 4.1 gives cut-off wavelengths for some common modes in rectangular waveguide and an equation for cut-off wavelength for any mode in rectangular guide.

Table 4.1. CUT-OFF WAVELENGTHS IN RECTANGULAR WAVEGUIDE

Mode	λ_c
$TE_{1,0}$	$2a$
$TE_{1,1}$ or $TM_{1,1}$	$\dfrac{2ab}{\sqrt{a^2 + b^2}}$
$TE_{2,0}$	a
$TE_{0,1}$	$2b$
$TE_{m,n}$ or $TM_{m,n}$	$\dfrac{2}{\sqrt{(m/a)^2 + (n/b)^2}}$

It is evident from Table 4.1 that for a particular size rectangular waveguide, the $TE_{1,0}$ mode has the highest cut-off wavelength of all modes. This means that for a given frequency, it is possible to choose waveguide dimensions so that only the $TE_{1,0}$ mode will propagate, and all other modes will have a given frequency below cut-off. For this reason, the $TE_{1,0}$ mode is called the *dominant mode* and all others are called *higher modes*.

When two or more modes propagate in a waveguide, they may interact in an undesirable manner. Thus, it is usually preferable to have only one mode at a time in a waveguide. Since the waveguide can be made to support only the dominant mode, that is the mode most commonly used. If a discontinuity in the guide causes a higher order mode to be excited, the new mode dies out quickly because it is below cut-off. However, if the waveguide were made to propagate the $TE_{2,0}$ mode, for example, a discontinuity might excite the $TE_{1,0}$ mode which would not die out. There are some special components and circuits which do utilize higher modes, but these are comparatively rare.

Table 4.2. CUT-OFF WAVELENGTHS IN CIRCULAR WAVEGUIDE

Mode	λ_c
$TE_{1,1}$	$1.706d$
$TM_{0,1}$	$1.306d$
$TE_{2,1}$	$1.028d$
$TE_{0,1}$	$0.820d$
$TM_{1,1}$	$0.820d$

In circular waveguide (as in rectangular), each mode has a cut-off wavelength which is a function only of the inside diameter, d, of the pipe. Table 4.2 gives cut-off wavelengths for some common modes in round waveguide. It is evident that the $TE_{1,1}$ mode is dominant and that a diameter can be chosen so that only this mode will propagate. However, because of the symmetry of the round waveguide and the unsymmetrical nature of the $TE_{1,1}$ mode, this mode presents difficulties and is not the most popular. It is evident that a bend or discontinuity might twist the mode in the pipe, leading to propagation in the wrong polarization.

From Fig. 4.4 it is evident that the $TM_{0,1}$ and $TE_{0,1}$ modes are symmetrical and would not be changed by twisting in the waveguide. Both of these modes have special uses. The $TM_{0,1}$ mode is used for rotary joints where its symmetry is important. This mode is chosen because of its high cut-off wavelength. The designer has to be careful to prevent the $TE_{1,1}$ mode from being excited, but doesn't have to worry about the others.

The $TE_{0,1}$ mode in circular waveguide is special in that it has lower attenuation than other modes and, of the five shown, is the only one in which the attenuation decreases as frequency increases. For this reason it is useful in long distance waveguide runs and at higher microwave frequencies. It does present problems to the designer who must prevent other modes from being excited.

4.3. GUIDE WAVELENGTH

It has been stated that the wavelength in the waveguide is different from the wavelength in air of free space. It can be shown that the relationship among *guide wavelength,* cut-off wavelength, and freespace wavelength is

$$\frac{1}{\lambda_g^2} = \frac{1}{\lambda_0^2} - \frac{1}{\lambda_c^2} \tag{4.1}$$

where λ_g is the guide wavelength, λ_0 is the free-space wavelength, and λ_c is the cut-off wavelength. Equation (4.1) may be solved for λ_g as follows:

$$\lambda_g = \frac{\lambda_0}{\sqrt{1 - (\lambda_0/\lambda_c)^2}} \tag{4.2}$$

This equation is true for any mode in waveguide of any cross section, provided the value of λ_c corresponds to the mode and cross section.

An examination of Eq. (4.2) indicates that if λ_0 is very much smaller than λ_c, the denominator is approximately unity, and the guide wavelength equals the free-space wavelength. As λ_0 approaches λ_c, λ_g increases and reaches infinity when λ_0 equals λ_c. When λ_0 exceeds λ_c, it is evident that λ_g is imaginary. This means that there can be no propagation in the guide when λ_0 is greater than the cut-off wavelength.

4.4. PHASE VELOCITY AND GROUP VELOCITY

Whenever a signal propagates in the waveguide, the guide wavelength is greater than the free-space wavelength (as can be seen in Eq. (4.2). Since the velocity of propagation is the product of the wavelength and the frequency, it follows that in a waveguide,

$$v_p = \lambda_g f \tag{4.3}$$

But the speed of light is equal to the product of the free-space wavelength and the frequency, as shown in Eq. (1.1). Thus, v_p is greater than the speed of light, since λ_g is greater than λ_0.

This seems to contradict the physical principle that no signal can be transmitted faster than the speed of light. However, the wavelength in the guide is the length of one cycle, and v_p represents the velocity of the phase. No intelligence or modulation travels at this velocity. Hence, v_p is called the *phase velocity*.

If there is modulation on the microwave carrier, the modulation envelope actually travels at a velocity slower than the carrier frequency and, of course, slower than the speed of light. The velocity of the modulation envelope is called the *group velocity* and is designated v_g. Thus, when a modulated signal travels in a waveguide, the modulation keeps slipping backward with respect to the carrier.

The phase velocity is greater than the speed of light by the ratio of λ_g to λ_0; thus,

$$v_p = \frac{\lambda_g}{\lambda_0} c \tag{4.4}$$

The group velocity is *shortened* by the same ratio.

$$v_g = \frac{\lambda_0}{\lambda_g} c \tag{4.5}$$

Multiplying these two equations together:

$$v_p v_g = c^2 \tag{4.6}$$

4.5. CHOICE OF DIMENSIONS

For the rectangular waveguide which is to operate in the $TE_{1,0}$ mode, dimensions are chosen so that only that mode will propagate. This means that the wide dimension must be more than a half-wavelength so that the free-space wavelength is less than λ_c—but this dimension should be less than one wavelength so that the $TE_{2,0}$ mode will not propagate. The narrow dimension should be less than a half-wavelength to prevent the $TE_{0,1}$ mode or others from being excited and propagating.

Similar considerations hold for round guides. The diameter is chosen so that only the dominant mode will propagate. If, however, the guide is used in applications requiring a higher mode, the diameter must be chosen to permit this mode to pass and, usually, to cut off still higher modes. The cut-off wavelengths shown in Tables 4.1 and 4.2 are a primary guide to choice of dimensions.

There are an infinite number of dimensions which will satisfy the conditions laid down by cut-off wavelengths. Another condition is imposed by attenuation. The larger the dimensions of the guide, the lower the attenuation. Thus, size is a compromise between attenuation and cut-off wavelength. A line of standard rectangular waveguide sizes has been introduced with microwaves. The narrow dimension is about half the width in these guides. However, in the early days of microwaves, outside dimensions were specified, but more recently (and more logically) inside dimensions have been used. Thus, an early waveguide size, still in use, has a half inch by one inch outside dimensions. Its wall thickness of 0.050 inch makes the inside dimensions 0.400 × 0.900 inch. A later standard size has inside dimensions of 0.622 × 0.311 inch, exactly a two-to-one ratio.

4.6. IMPEDANCE

Impedance may be calculated from the relationship involving power and voltage, power and current, or voltage and current. Thus,

$$Z = \frac{E}{I} \tag{4.7}$$

$$Z = \frac{P}{I^2} \tag{4.8}$$

$$Z = \frac{E^2}{P} \tag{4.9}$$

The reason for the different values of Z is that the current and voltage are not uniquely defined in a waveguide as they are in lumped circuits, where all three equations give the same value for Z. Ordinarily, the difference in values doesn't matter, since in joining two waveguides the mismatch would be the same no matter which relationship is chosen. However, in joining a waveguide to a coaxial line [Eq. (4.9)], the power-voltage relationship, most nearly approaches the characteristic impedance concept of the coaxial line. Using this, the characteristic impedance Z_0 of a waveguide is

$$Z_0 = 377 \frac{b}{a} \frac{\lambda_g}{\lambda_0} \tag{4.10}$$

for all TE modes, and

$$Z_0 = 377 \frac{b}{a} \frac{\lambda_0}{\lambda_g} \tag{4.11}$$

for all TM modes.

It should be noticed that the impedance is not constant with frequency. The wide and narrow dimensions of the waveguide, are a and b respectively. For round guides, it may be assumed that $a = b$.

4.7. DIELECTRIC

In all previous equations and relationships it has been assumed that the waveguide was empty or was filled with air, which has a relative dielectric constant close to unity. However, in many applications, the waveguide might be filled with a low-loss dielectric material in order to reduce the size of the system. The equations and quantities discussed in this chapter are affected by the relative dielectric constant of the material.

For simplicity it has been assumed in other equations that the cut-off wavelengths indicated in Tables 4.1 and 4.2 are unchanged when the waveguide is loaded with a dielectric. However, to determine if a frequency is below cut-off, its corresponding free-space wavelength must first be divided by the square root of the dielectric constant. Thus, suppose that "X-band" waveguide, which has a wide dimension of 0.900 inch is filled with Polystyrene which has a dielectric constant $\epsilon = 2.56$. $\lambda_c = 1.8$ inches. The waveguide, if empty, would pass 6600 megacycles ($\lambda_0 = 1.78$ inches) but would stop 6500 megacycles ($\lambda_0 = 1.82$ inches). However, with the dielectric, 4000 megacycles will propagate; although its wavelength is 2.95 inches,

$$\frac{2.95}{\sqrt{2.56}} = \frac{2.95}{1.6} = 1.71 \text{ in.},$$

which is less than λ_c. The guide wavelength is shortened when the guide is loaded. The new equation is

$$\lambda_g = \frac{\lambda_0}{\sqrt{\epsilon - (\lambda_0/\lambda_c)^2}} \tag{4.12}$$

Equation (4.12) should be used all the time instead of Eq. (4.2). When the dielectric is air or a vacuum, ϵ is unity, and Eq. (4.12) reduces to Eq. (4.2).

The complete equation for impedance of TE modes in a waveguide is

$$Z_0 = 377 \sqrt{\frac{\mu}{\epsilon}} \frac{b}{a} \frac{\lambda_g}{\lambda_0} \tag{4.13}$$

The relative permeability μ is unity for nonmagnetic metals and thus can usually be omitted. The dielectric constant has already affected λ_g in Eq. (4.13) and it also appears again. For TM modes, the reciprocal of λ_g/λ_0 is used as in Eq. (4.11).

The phase velocity is still as given by Eq. (4.3), $v_p = \lambda_g f$, but uses the value of λ_g determined by Eq. (4.12). However, in comparing this to the speed of light, as in Eq. (4.4), the wavelength in free space is changed to the wavelength in a dielectric-filled free space. Thus, for a waveguide filled with a dielectric, instead of using Eq. (4.4), the phase velocity becomes

$$v_p = \frac{\lambda_g}{\lambda_d} c \tag{4.14}$$

where

$$\lambda_d = \frac{\lambda_0}{\sqrt{\epsilon}} \tag{4.15}$$

Similarly, the group velocity is

$$v_p = \frac{\lambda_d}{\lambda_g} c \tag{4.16}$$

and, of course, Eq. (4.6) still holds.

4.8. DISCONTINUITIES

A piece of metal or dielectric partially filling the waveguide introduces a discontinuity which will reflect part of the incident microwave signal. It should be evident that if a metal obstacle such as a piece of wire is parallel to the electric field, it will have more effect than if it is perpendicular. For example, if the obstacle extends from top to bottom of the waveguide, as in Fig. 4.5a, it should act as a short circuit for some of the voltage lines, but if it is across the waveguide as in Fig. 4.5b, it will have a negligible shorting effect and consequently cause negligible reflection. It should be obvious, also, that when the wire is across the center of the waveguide where the voltage is a maximum, as in Fig. 4.5a, it will cause a greater reflection than when it is nearer a side wall, as in Fig. 4.5c. Similarly, a dielectric discontinuity will have maximum effect when it intercepts a maximum voltage field. Although not classed as a discontinuity, resistive material will interact in the same manner; that is, a thin resistive sheet parallel to the voltage lines will absorb more

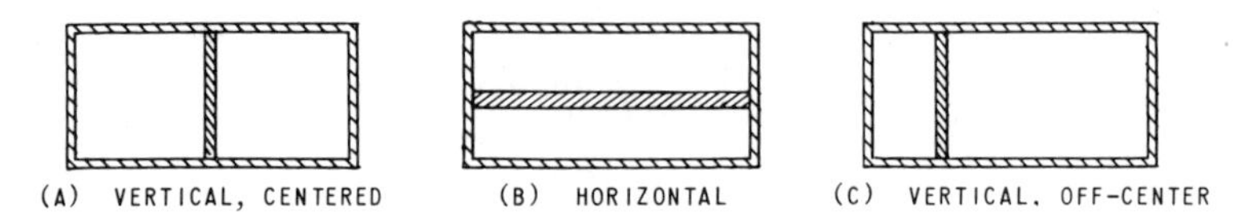

Fig. 4.5. Thin wire or post.

power than one perpendicular to these lines and will be more effective in the center than nearer the side walls of the waveguide.

Thin discontinuities may be considered to be lumped circuit elements in shunt with the transmission line and are classed as either inductive or capacitive. Because the discontinuities do have finite thickness there is also some series reactance associated with them, but in practice, discontinuities are usually thin so that the series reactance has a negligible effect compared to the shunt susceptance.

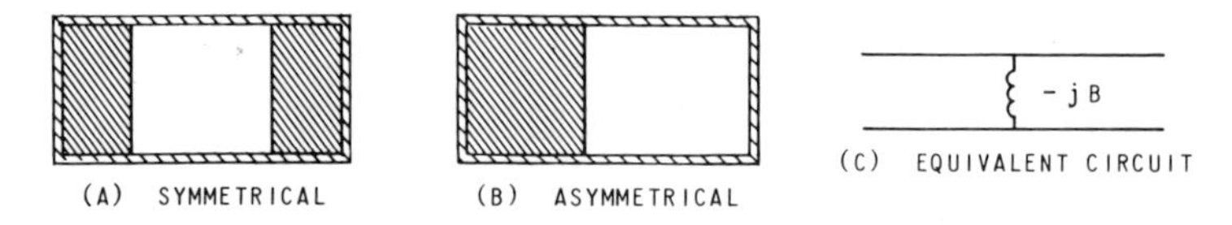

Fig. 4.6. Inductive irises.

Irises are thin metal discontinuities partially closing the waveguide. Figure 4.6 shows the two types of inductive irises, symmetrical and asymmetrical, and their equivalent circuit, an inductance susceptance shunting a transmission line. The opening of the iris looks like a waveguide below cut-off, but because the iris is thin, there is some propagation through the opening and, of course, some reflection. For both the symmetrical and the asymmetrical cases, the smaller the opening, the larger is the shunt susceptance. It should be noted that if the iris is followed by a matched line, the VSWR at the iris is $1 - jB/Y_0$, since the iris is in parallel with a line whose normalized characteristic admittance is unity.

A post across the waveguide, as shown in Figs. 4.5a or 4.5c, is also an inductive shunt susceptance and has the equivalent circuit of Fig. 4.6c.

Since posts are more easily machined than irises, they are more commonly used. However, a single post cannot have too great a diameter or its series reactance may become appreciable. Obviously, the nearer to the center the post is situated, the greater will be the shunt susceptance. It is possible to use two posts in the same plane, symmetrically spaced, to get larger values of inductive susceptance than are possible with a single post at the center.

Capacitive irises are shown in Fig. 4.7. These also may be symmetrical or asymmetrical. As before, the equivalent circuit is a shunt susceptance

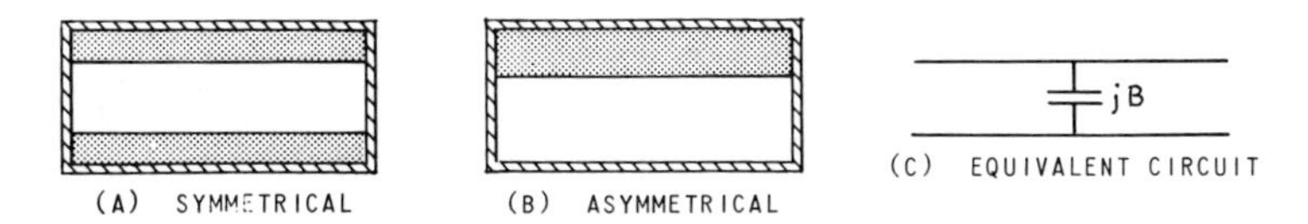

Fig. 4.7. Capacitive irises.

across the transmission line, but in this case the susceptance is capacitive. Again, as the opening is made smaller, the susceptance is increased. Capacitive irises should be avoided at high power levels because the reduced gap may cause voltage breakdown.

Because of the ease of fabrication and the ability to adjust the discontinuity, a screw projecting into the waveguide is one of the most popular types of discontinuities. (A cross section is shown in Fig. 4.8.) It is not necessary for the screw to be located on the center line of the waveguide. When the screw is first inserted, it presents a capacitive discontinuity and may be represented approximately by the equivalent circuit of Fig. 4.7c. Obviously, when it is all the way across the waveguide, it looks like an inductive post, and thus there must be a transition point. This occurs when the screw projects about five-eighths of the way across the waveguide, at which point it has infinite susceptance. Thus as the screw is inserted, it begins as a capacitive susceptance, whose value increases to infinity at the transition point, and then decreases as an inductive susceptance.

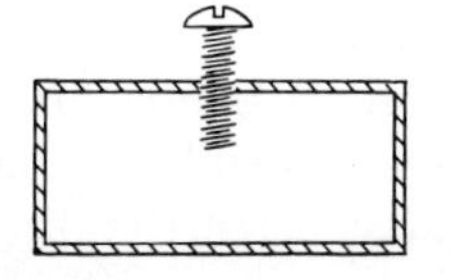

Fig. 4.8. Tuning screw.

All of these discontinuities can also be used in circular guide. In this case there are no wide and narrow dimensions, but the test whether a discontinuity is inductive or capacitive is really the same as in rectangular guide. If the edge of the discontinuity is parallel to the voltage lines, it is inductive; if it reduces the gap between the points of voltage difference, it is capacitive.

4.9. BENDS, CORNERS, AND TWISTS

When it is necessary to change the direction of a waveguide in order to get around obstacles or to join waveguide runs in a system, it is permissible to bend the waveguide in either plane as shown in Fig. 4.9. The bend is designated an *E-plane* bend if the voltage line in the cross section changes direction. Figure 4.9a shows an E-plane bend and the voltage vectors at the faces; Fig. 4.9b shows a bend in which the voltage vectors always have the same direction, but the magnetic lines do not. This is therefore an *H-plane* bend. There is a simple mnemonic aid to differentiate

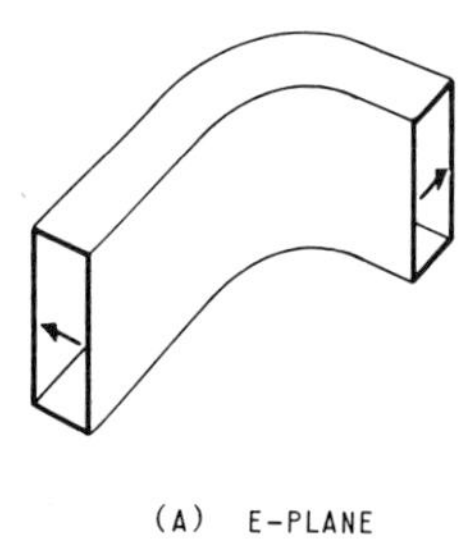

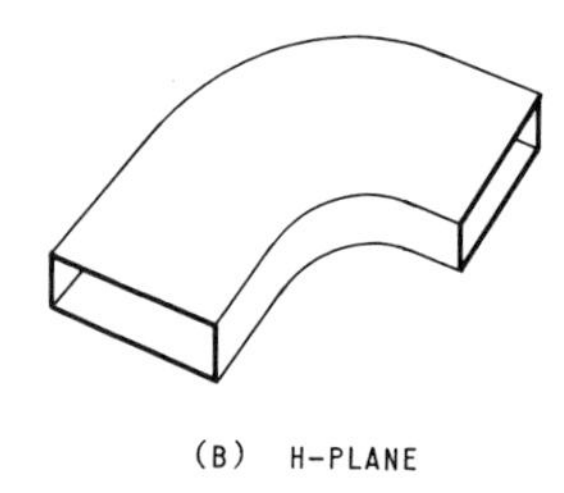

Fig. 4.9. Waveguide bends.

between the two. If one tries to bend a guide it is obviously more difficult to form the H-plane bend than the E-plane. Thus, H-plane is the Hard bend and E-plane is the Easy. However, it is better to learn the exact meanings so that bends in square waveguide can be described.

The cross section of the bent waveguide should be uniform to prevent reflections. In the early days of microwaves, in order to minimize the possibility of high reflections, bends were long gradual curves many wavelengths long. However, with precision casting techniques it has become possible to make low VSWR bends, in either plane, with radii no larger than the dimension that is bent. These precision bends can be used over the entire frequency band for which the waveguide is designed.

Bends in rectangular waveguide, even when well machined, do occasion twisting of the dominant mode and generation of higher modes. However, the dominant mode can only be supported in one polarization in the waveguide and straightens out after the bend. The higher modes also die out quickly because they are below cut-off.

In general, bends are undesirable in circular waveguides. If the dominant mode is twisted, it will propagate in whatever polarization it finds itself. The amount of twisting is not controllable so that it may be impossible to transmit correctly.

When a bend takes up too much room, some space may be saved by substituting a double-mitred corner, as shown in Fig. 4.10. This type of joint can be made in either the broad wall (H-plane) or narrow wall

(E-plane) of the waveguide. In effect, there are two discontinuities spaced approximately a quarter of a guide wavelength apart so that their reflections will cancel. Because of this, this type of bend or corner is frequency sensitive and cannot be used in broad-band applications.

Smooth bends may be made with the arms coming off at any angle. The angle of the bend is designated by the difference between the angle that the arms make and a straight angle. Thus, in a 70° bend, the arms are 110° apart.

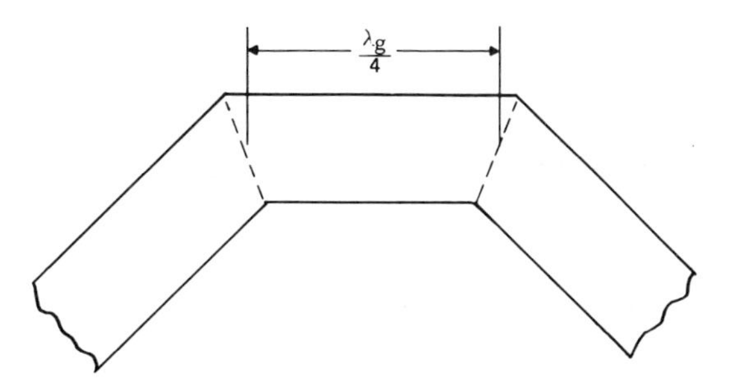

Fig. 4.10. Double-mitred corner.

Double-mitred corners can be used for bends up to 90°; for tighter bends, triple or quadruple corners are used. The angle is designating as with smooth bends. In constructing a double-mitred corner, each mitred joint changes direction by half the desired angle.

Rectangular waveguide may also be twisted to change the direction of polarization. Twists should be long in comparison to a wavelength in order to minimize reflections. For a typical example, a 90° twist should be at least four guide wavelengths long. For smaller angles, shorter lengths may be used. The important thing is to maintain a nearly uniform cross section throughout the twist.

4.10. MODE FILTERS

In Sec. 4.2 it was indicated that there are applications where it is desirable to use a mode in a waveguide other than the dominant mode. However, when a higher mode is propagating in a waveguide, any discontinuity or asymmetry may induce the dominant mode or, in fact, any lower modes to propagate. Unlike the cut-off phenomenon when the tables are reversed, there is no mechanism to stop the dominant or lower mode once it is started. Since the lower modes derive their energy from the wanted mode, the latter is degraded whenever it loses energy to other modes. To prevent lower modes from forming, mode filters are used near every discontinuity that might start an unwanted mode.

Basically, a mode filter is a discontinuity that takes advantage of the different field patterns of the various modes. For the unwanted mode, it is placed at a point of high voltage; for the desired mode, it is set at low voltage or zero. Thus, if a thin wire is placed across the center of a rectangular waveguide as in Fig. 4.5a, it is at a point of maximum voltage for the dominant mode, as can be seen by the end view of Fig. 4.2. However, for the $TE_{2,0}$ mode, as is evident from Fig. 4.3, the wire is at a point of zero voltage. Thus, the wire looks like a large inductive susceptance to the $TE_{1,0}$ mode, causing it to be reflected. At the same time it has no effect on the $TE_{2,0}$ mode which can propagate past it as if there were no obstruction.

The types and varieties of mode filters are too numerous to describe since they can be used in any combination of wanted and unwanted modes. The designer need only look for the place in the waveguide where some modes can be reflected without disturbing the required mode.

4.11. POWER HANDLING CAPABILITY

When a signal propagates in a waveguide, a voltage exists between the walls of the guide, as shown by the voltage lines in Figs. 4.3 and 4.4. The maximum power that can be transmitted in a waveguide is limited by voltage breakdown. At sea level and normal temperature and humidity, experiment has shown that air breaks down at a field strength of approximately 30,000 volts per centimeter. This value may be used to calculate the breakdown or maximum power of a waveguide.

For a rectangular waveguide operating in the dominant mode, the maximum power is

$$P = 600{,}000ab\left(\frac{\lambda}{\lambda_g}\right) \text{ w} \tag{4.17}$$

where a and b are the waveguide dimensions in centimeters. Thus, X-band waveguide, which has inside dimensions 0.9 by 0.4 inches, or 2.286 by 1.016 centimeters, has a breakdown power of

$$600{,}000 \times 2.286 \times 1.016 \times 0.75 = 1.04 \text{ Mw}$$

at the frequency where λ/λ_g is 0.75. This value of λ/λ_g is near the middle of the usable waveguide frequency band for all waveguides.

For a circular waveguide operating in the dominant or $TE_{1,1}$ mode, the maximum power is

$$P = 450{,}000d^2 \frac{\lambda}{\lambda_g} \text{ w} \tag{4.18}$$

where d is the diameter of the round guide in centimeters.

If the waveguide is pressurized above normal atmospheric pressure, it can accommodate higher powers without breakdown. Conversely, if the pressure is decreased, as when the radar set is airborne, the power capability is reduced. Of course, in a perfect vacuum there would be no breakdown at all, because there would be no gas to ionize.

If there is a discontinuity in the waveguide, the reflected voltage will add to the incident and cause breakdown at a lower power level than if there were no discontinuity. This plus variations in breakdown caused by dust particles, sharp points or corners, and such ambient conditions as temperature and humidity make it undesirable to operate the waveguide too close to its rated power level. In practice it is customary to allow a factor of approximately four for safety. Thus the X-band waveguide described would not be used above a quarter of a megawatt without additional precautions.

To raise its power handling capability, a waveguide can be pressurized as was indicated above. It can also be filled with a material having a higher breakdown field strength than air. Such materials include gases, like sulfur hexofluoride, and foamed dielectrics, neither of which will cause appreciable change in the transmission characteristics of the waveguide. Of course, if the waveguide is dielectric-loaded it will stand much higher fields.

4.12. ATTENUATION

Although transmission lines are frequently treated as if they were lossless, they do have some finite attenuation. Losses in a waveguide come from two sources, conductor losses and dielectric losses. If the walls of the waveguide were perfect conductors with infinite conductivity, there would be no conductor loss, but no material exists to permit this.

The conductor loss (also called copper loss and I^2R loss) in a waveguide depends on the skin depth, as mentioned in Sec. 2.2, and thus is a function of frequency [see Eq. (2.1)]. For a copper rectangular waveguide operating in the dominant mode, the attenuation is

$$\alpha = \frac{0.01107}{a^{3/2}}\left[\frac{\frac{a}{2b}\left(\frac{f}{f_c}\right)^{3/2} + \left(\frac{f_c}{f}\right)^{1/2}}{\sqrt{\left(\frac{f}{f_c}\right)^2 - 1}}\right] \tag{4.19}$$

where α is in decibels per foot, a and b are the dimensions of the guide in inches, f_c is the cut-off frequency, and f is the frequency of operation. If the material is a metal other than copper, Eq. (4.19) must be multiplied by the square root of the ratio of the resistivity of the metal to that of copper.

Equation (4.19) indicates that when the frequency approaches the cut-off frequency, the attenuation approaches infinity. However, although the attenuation is very high near cut-off, this equation is not accurate near cut-off. It is accurate in the frequency band at which a waveguide is normally used. The quantity $a^{3/2}$ in the denominator indicates that larger waveguides have lower attenuation. Since larger waveguides are used at lower frequencies, it is evident that for very high frequencies using very small waveguide, attenuation is a problem. As an example, in 1.340×2.840 inch copper waveguide used at 3000 megacycles, the attenuation is about 0.006 decibels per foot, an almost negligible amount. But at 25,000 megacycles, the standard waveguide is 0.420×0.170 inch, which has an attenuation (in copper) of about 0.1 decibel per foot, which is appreciable in long waveguide runs. Other modes in rectangular waveguide are attenuated more than the $TE_{1,0}$ mode.

In circular waveguide, the attenuation for the three most used modes is given by the following equations:
For the $TE_{1,1}$ mode,

$$\alpha = \frac{0.00423}{a^{3/2}} \frac{(f_c/f)^{1/2} + 1/2.38(f/f_c)^{3/2}}{\sqrt{(f/f_c)^2 - 1}} \tag{4.20}$$

For the $TM_{0,1}$ mode,

$$\alpha = \frac{0.00485}{a^{3/2}} \frac{(f/f_c)^{3/2}}{\sqrt{(f/f_c)^2 - 1}} \tag{4.21}$$

For the $TE_{0,1}$ mode,

$$\alpha = \frac{0.00611}{a^{3/2}} \frac{(f_c/f)^{1/2}}{\sqrt{(f/f_c)^2 - 1}} \tag{4.22}$$

In these three formulas, α is in decibels per foot and a is the radius of the round waveguide in inches. If a material other than copper is used, the values of α given by Eqs. (4.20), (4.21), and (4.22), should be multiplied by the square root of the ratio of the resistivities.

It is interesting to note that, if the guide diameter is fixed, the attenuation of the $TE_{0,1}$ mode decreases as the frequency is increased. For this reason, this mode is used at millimeter wavelengths where rectangular waveguide attenuation is prohibitive. It is also used in long waveguide runs for microwave communications. With the $TE_{0,1}$ oversize guide is used and care is taken to prevent other modes from being excited.

4.13. MATERIALS

Waveguides are usually made of metal. The choice of material is frequently the result of a compromise among such factors as low resistivity, ease of fabrication, weight, corrosion resistance, etc. In the laboratory, where changes are made after every test, it is usually desirable to use a material that can be soldered and machined with ease. In this case brass is used. For airborne applications, low weight is the most important requirement, so the choice is aluminum or magnesium. When design work is complete, precision casting may be used to make the component in quantity. Bronze and aluminum are both good casting materials, and the latter, of course, is used whenever weight is important.

In systems where every fraction of a decibel is important, the conductor losses in brass and aluminum are too great. Thus, waveguide components made of these metals are usually silver-plated. The aluminum guide is first given a copper flash, since silver doesn't plate readily on aluminum. The depth of the silver plating is more than the skin depth in silver so that there is no current flowing in the base metal. At millimeter wavelengths, the waveguide is so small that it is usually cheaper to make the guide entirely of silver rather than silver-plate some other metal. Gold plate is also a good conductor, but is more expensive than silver. It has an additional advantage that it is corrosion-resistant, whereas silver is not.

To prevent silver surfaces from tarnishing or corroding, the surface is plated with a very thin layer of rhodium. Rhodium will not corrode, but it has a higher resistivity than silver. However, by making the thickness of the rhodium plating much less than the skin depth, most of the current will flow in the silver underneath it.

4.14. FLANGES

A complicated microwave system is comprised of many individual components which are designed and developed individually and are then connected together. In long waveguide runs it is usually desirable to make up the waveguide in shorter sections to facilitate replacement or repair. Connections are usually made by means of flanges fastened to the ends of the waveguide sections. There are two kinds of flanges in general use. *Cover flanges* are flat metal surfaces perpendicular to the direction of propagation in the guide. Two cover flanges may be joined by bolts so that the inside of the waveguide is a continuous surface, as in Fig. 4.11. In high power applications it is important that the cover flanges are well lapped and fastened accurately, since any discontinuity in the inside surface of the waveguide will cause voltage breakdown.

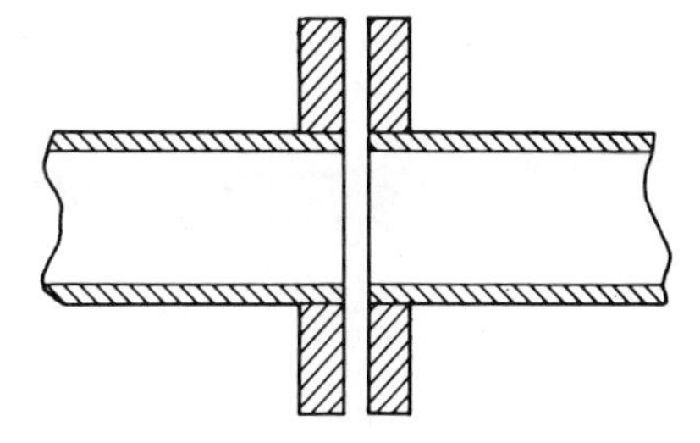

Fig. 4.11. Waveguide cover flanges.

When the guides do not mate accurately, a *choke joint* is generally used to insure a good electrical connection. This is a connection using a *choke flange* and a cover flange as in Fig. 4.12. The choke flange has a circular slot which is effectively a short-circuited line a quarter-wavelength long. Thus, at A, around the waveguide, it looks like an open circuit or infinite impedance. The slot is also a quarter-wavelength from the waveguide so that, at the waveguide, the infinite impedance is transformed back to a short circuit. The waveguides, consequently, appear to be connected even

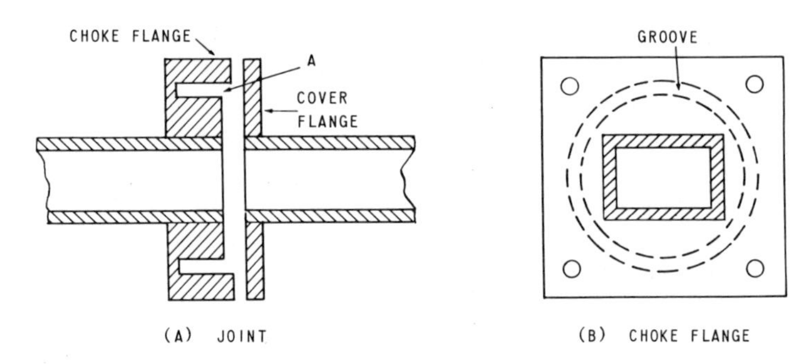

Fig. 4.12. Choke joint.

when they are physically apart or twisted slightly with respect to one another.

The choke joint is also useful when it is necessary to separate the two waveguides but maintain an r-f connection. For example, if a d-c voltage is applied to one section of waveguide, while other sections are at ground, choke joints separated by a thin piece of insulating material will prevent a d-c short and yet maintain an r-f connection. Choke joints are also used when it is desirable to prevent mechanical vibrations in one part of a system from reaching a more delicate component.

4.15. RIDGE WAVEGUIDE

Rectangular and circular waveguide have limited frequency bandwidths where they can be used without danger of propagating higher modes. Thus, a useful bandwidth in rectangular waveguide is approximately one and a half to one. The low frequency limit must be sufficiently above the cut-off frequency so that attenuation is not too severe, while the high frequency limit must be below cut-off for the $TE_{2,0}$ mode. Circular waveguide frequency bands are even more restricted, since cut-off wavelengths for the $TE_{1,1}$ mode and the $TM_{0,1}$ mode are closer together.

The bandwidth can be increased by using ridge guide—illustrated in

Figs. 2.3d and 2.3e. The ridge acts as a capacitive loading in the waveguide and thus lowers both the characteristic impedance and the phase velocity. The cut-off frequency for the $TE_{1,0}$ mode is consequently lowered, sometimes by as much as a factor of five. At the same time, by proper design of the ridge, the cut-off frequencies for the $TE_{2,0}$ and $TE_{3,0}$ modes may be increased. Thus, the bandwidth is increased; however, at the same time copper losses are also increased and power handling capacity is reduced.

QUESTIONS AND PROBLEMS

4.1. Describe the method of designating the mode of transmission in a rectangular guide.

4.2. Why is transmission in the dominant mode most often used in transmission for rectangular but not for circular guides?

4.3. Calculate the cut-off frequency for $TE_{1,0}$, $TE_{1,1}$, and $TM_{1,1}$ modes for a rectangular guide 1 cm × 2 cm.

4.4. What is peculiar about the $TE_{0,1}$ mode in a circular guide?

4.5. Calculate the guide wavelengths of the sections shown at 10 Gc.

ans. a. 1.47 in.

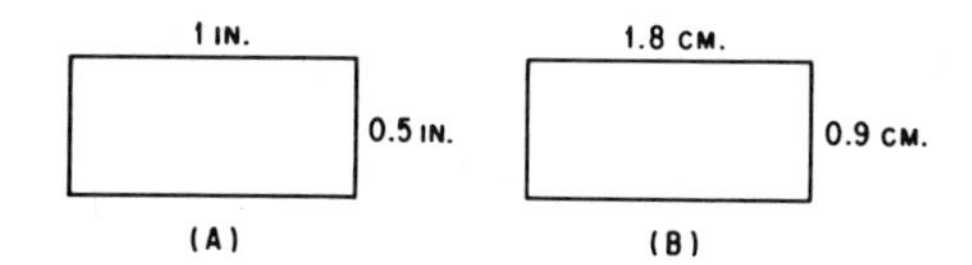

Fig. P 4.5

4.6. The dimensions of a guide are 2.5 cm × 1 cm. The frequency is 8.6 Gc. Find the following.
a. Possible modes.
b. Group velocity.
c. Phase velocity.
d. Cut-off frequency.
e. Guide wavelength.
f. Characteristic impedance [use Eq. (4.10)].

4.7. Repeat Prob. 4.6a but assume that the guide is filled with Polystyrene.

4.8. Explain the action of the screw tuner shown in Fig. 4.8.

4.9. What determines whether a bend is of the "H" or "E" type?

4.10. Why are bends more tolerable in rectangular than in circular guide?

4.11. What factors determine the need for and the placement of a mode filter?

4.12. What is the maximum power (at atmospheric pressure) that can be transmitted by rectangular guide 1 cm $\times$ 2 cm at 9 Gc? (1Gc = 1000 megacycles) What factors make it necessary to lower this figure in practice?

4.13. Discuss the factors that determine the amount of attenuation in a waveguide. What are some of the means taken to minimize the amount of attenuation?

4.14. Calculate the approximate db loss per ft for the conditions (copper guide) of Prob. 7.6.

4.15. Describe two types of joints for rectangular guides and give the advantages of each.

5

COAXIAL LINES

Coaxial lines are transmission lines which have two conductors, one inside the other. (A cross section is shown in Fig. 2.2a.) The engineer who works at low frequencies has no difficulty understanding the coaxial line since the two conductor system is used at low frequencies too. However, at microwaves, where the lengths of line may be several wavelengths, the coaxial line should be considered as a transmission line in which electromagnetic waves are transmitted in the medium between the two coaxial cylinders. This medium may be air or dielectric.

5.1. MODES

The principle mode in the coaxial line is shown in Fig. 5.1. The voltage lines extend from one conductor to the other and thus have no component in the direction of propagation. The magnetic lines are closed loops and also have no component in the direction of propagation. The mode is both transverse electric and transverse magnetic. It is designated the *TEM mode*. It has no cut-off wavelength and can propagate at all frequencies from direct current through microwaves. In practice, the upper limit is the frequency at which higher modes can propagate.

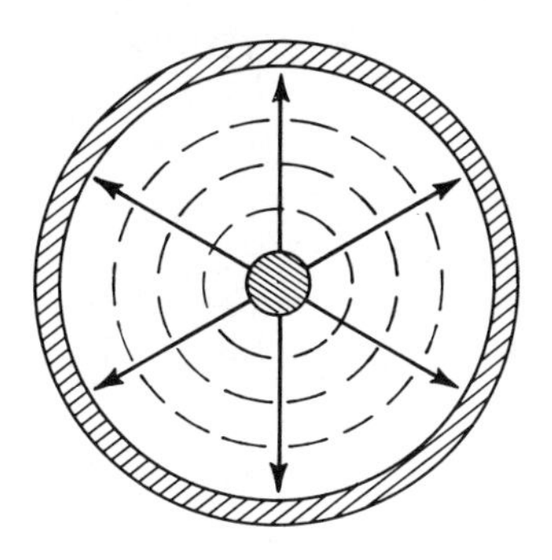

Fig. 5.1. TEM mode.

Many of the modes of circular waveguide can exist in coaxial line with some distortion caused by the inner conductor. The first higher mode in coaxial line, shown in Fig. 5.2, is the $TE_{1,1}$ mode, or dominant mode of circular waveguide. However, the cut-off wavelength for this mode in coaxial line is not the same as it is in round waveguide. An approximate value of cut-off may be determined by considering the $TE_{1,0}$ mode in coaxial line as being developed from the $TE_{2,0}$ mode in rectangular

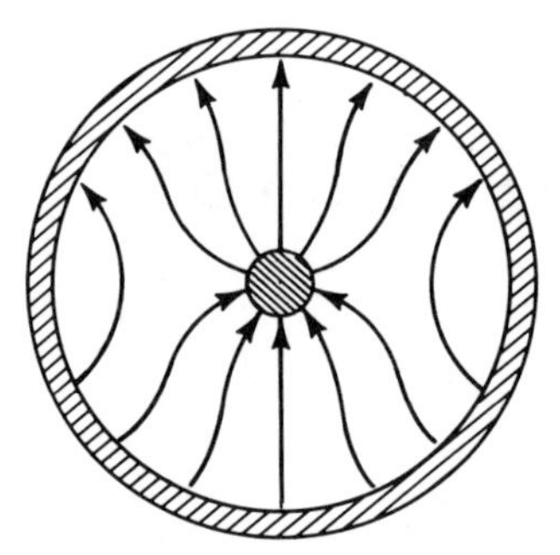

Fig. 5.2. $TE_{1,1}$ mode in coaxial line.

waveguide as shown in Fig. 5.3. The two end walls of the rectangular waveguide are bent around until they touch. Then, one broad wall becomes the inner conductor and the other the outer conductor of the coaxial line. The resultant configuration of the original $TE_{2,0}$ mode is the $TE_{1,1}$ mode in the coaxial line. It may be assumed that the cut-off wavelength for the new $TE_{1,1}$ mode is the same as the cut-off wavelength for

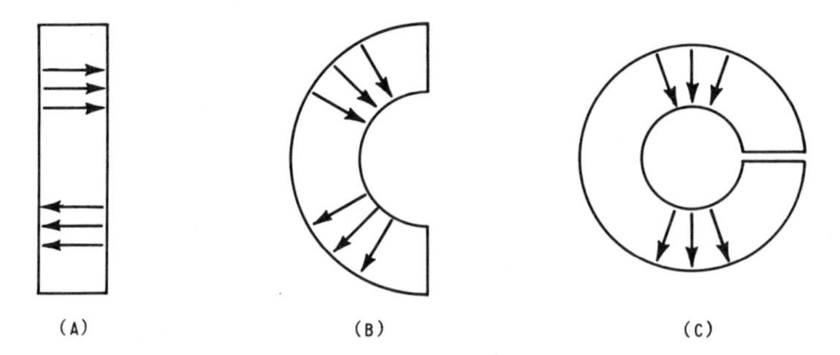

Fig. 5.3. Development of $TE_{1,1}$, mode in coaxial line.

the original $TE_{2,0}$ mode, which was the dimension of the broad wall of the waveguide. Since one broad wall has been stretched to be the circumference of the outer conductor of the coaxial line while the other is the circumference of the inner conductor, an average of the two values is a close enough approximation to the cut-off wavelength.

$$\lambda_c \approx (a + b)\pi \tag{5.1}$$

where a is the outside radius of the inner conductor and b is the inside radius of the outer.

5.2. CHARACTERISTIC IMPEDANCE

A unit section of coaxial line has inductance, capacitance, and resistance. The inductance per unit length in microhenries per centimeter is

$$L = 0.0046 \log_{10} \frac{b}{a} \tag{5.2}$$

The capacitance per unit length in picofarads per centimeter is

$$C = \frac{0.241\epsilon}{\log_{10} (b/a)} \tag{5.3}$$

where ϵ is the relative dielectric constant of the medium between the conductors. The resistances per unit length for copper conductors in ohms per centimeter is

$$R = 4.14 \times 10^{-8} \sqrt{f} \left(\frac{1}{b} + \frac{1}{a}\right) \tag{5.4}$$

where a and b are measured in centimeters. The resistance varies with frequency because of the skin effect. If the coaxial line is made of a nonmagnetic material other than copper, the resistance in Eq. (5.4) should be multiplied by the square root of the ratio of the resistivity of the new material to that of copper.

The characteristic impedance of a low-loss coaxial line is

$$Z_0 = \sqrt{\frac{L}{C}} = \frac{138}{\sqrt{\epsilon}} \log_{10} \frac{b}{a} \tag{5.5}$$

Eq. (5.5) indicates that the characteristic impedance does not vary with frequency.

5.3. PHASE VELOCITY

The wavelength in a coaxial line is the same as the wavelength in an unbounded medium that has the same dielectric constant; that is, in a coaxial line,

$$\lambda_g = \lambda_d = \frac{\lambda_0}{\sqrt{\epsilon}} \tag{5.6}$$

Thus the symbol λ_g is not used for coaxial lines. The velocity of propagation is then

$$v = \frac{c}{\sqrt{\epsilon}} \tag{5.7}$$

where c is the velocity in free space. If the dielectric in the coaxial line is air, $\epsilon = 1$, and the velocity of propagation is the same as the speed of light. There is no distinction between phase velocity and group velocity as there is in waveguides.

5.4. ATTENUATION

The attenuation in a coaxial line, as in other forms of transmission line, comes from two sources of loss. These are the copper loss in the conductors and the dielectric loss in the medium between the conductors. A third source of attenuation, radiation loss, is of no importance, because the coaxial line is completely shielded.

The attenuation caused by conductor losses in a coaxial line is

$$\alpha_c = 13.6 \frac{\delta}{\lambda b}\left(1 + \frac{b}{a}\right)\frac{\sqrt{\epsilon}}{\ln (b/a)} \tag{5.8}$$

where α_c is the attenuation in decibels per unit length, δ is the skin depth, λ is the wavelength, b is the outer radius, a is the inner radius, and ϵ is the relative dielectric constant of the medium between the conductors. In Eq. (5.8), δ, λ, b, and a must all be in the same unit of length.

Assuming that the outer dimension, b, of the coaxial line remains fixed, Eq. (5.8) indicates that the minimum attenuation occurs when $b/a = 3.6$. With air dielectric, this corresponds to a characteristic impedance of 77 ohms.

When the space between the conductors is filled with a dielectric other than air, there are additional losses. The dielectric may be assumed to have a complex dielectric constant.

$$\epsilon = \epsilon' - j\epsilon'' \tag{5.9}$$

where ϵ' is the real part, and ϵ'' is the imaginary part of ϵ. The *loss tangent* of the dielectric is

$$\tan \delta = \frac{\epsilon''}{\epsilon'} \tag{5.10}$$

For practical dielectrics, $\tan \delta$ varies from 0.0001 to 0.01. Thus, ϵ'' is very small compared to ϵ'. Sometimes the loss in a dielectric is described by its *power factor,* which is actually the power factor of a condenser which contains the dielectric. The power factor is exactly

$$\text{pf} = \cos (90° - \delta) = \frac{\epsilon''}{\sqrt{(\epsilon')^2 + (\epsilon'')^2}} \tag{5.11}$$

but when ϵ'' is much less than ϵ', the power factor is very nearly equal to the loss tangent.

The attenuation in decibels per unit length in a coaxial line caused by losses in the dielectric is

$$\alpha_d = 27.3 \frac{\sqrt{\epsilon'} \tan \delta}{\lambda} \tag{5.12}$$

From Eq. (5.12) it is evident that the dielectric loss increases with frequency.

The total attenuation in a coaxial line is simply the sum of Eqs. (5.8) and (5.12). Thus,

$$\alpha_T = \alpha_c + \alpha_d \tag{5.13}$$

5.5. VOLTAGE BREAKDOWN AND MAXIMUM POWER

The electric field intensity at any point between the inner and outer conductors of a coaxial line is

$$E = \frac{V}{r \ln b/a} \tag{5.14}$$

where E is measured in volts per centimeter, V is in volts, and r is the radius of the point in centimeters. The radius, r, varies from a to b. The intensity, E, is thus a maximum for a given line when $r = a$; that is, when measured at the surface of the inner conductor. For a given value of b, the maximum voltage between conductors before breakdown occurs when b/a equals 2.718. With air dielectric this corresponds to a characteristic impedance of 60 ohms.

The maximum power that can be carried in a coaxial line doesn't occur at the same impedance as the maximum voltage. At a fixed value of b, $P = V^2/Z_0$, and consequently a lower value of Z_0, even if V is also decreased, might yield higher power. Maximum power can be carried when b/a is 1.65, which corresponds to an air-filled coaxial line with characteristic impedance of 30 ohms.

5.6. FIXED FREQUENCY

In the two preceding sections, the optimum characteristic impedances were presented on the assumption that the outer conductor was fixed. However, if b can be varied while the frequency of operation is fixed, other values of characteristic impedance will give maximum power-carrying capacity and minimum attenuation. This occurs because of the limitation that a and b must be so chosen that the higher $TE_{1,1}$ mode will not propagate.

Within this limitation, at a fixed frequency, the maximum power-carrying capacity occurs at a characteristic impedance of 44 ohms. The minimum attenuation occurs at 93 ohms.

5.7. CHOICE OF DIMENSIONS

Thus far, depending upon the conditions of the problem and what quantity is to be maximized, five different values of Z_0 have been presented as optimum. These values have ranged from 30 to 93 ohms. Although in theory a particular line could be designed specifically for the particular application in which it is to be used, this would present difficult problems of measurement. In order to standardize measuring equipment and also to have a standard assortment of commercially available coaxial lines, it is necessary to compromise and use lines which do not have optimum values of characteristic impedance. A satisfactory value is 50 ohms. As a matter of fact, the variation in the several quantities mentioned above with characteristic impedance is slow, so that at 50 ohms the values have not decreased too far from their maxima. Table 5.1 lists the value of characteristic impedance which yields the best value of the required quantity. In the second column, the efficiency of a 50-ohm line is compared to the optimum line.

Table 5.1. OPTIMUM CHARACTERISTIC IMPEDANCE

Conditions	Optimum Z_0	Efficiency of 50 ohm line
Maximum power-carrying capacity with fixed wavelength	44 ohms	99 per cent
Maximum power-carrying capacity with fixed outer conductor	30	86
Minimum attenuation with fixed wavelength	93	78
Minimum attenuation with fixed outer conductor	77	91
Maximum breakdown voltage with fixed outer conductor	60	99

5.8. STUB SUPPORTS

The inner conductor of a coaxial line must be supported on the center line of the outer conductor. One way of accomplishing this is by using short-circuited stubs as supports. A short-circuited line which is one quarter-wavelength long has an infinite input impedance, so that at the inner conductor there is apparently no connection, even though it is shorted for direct current. However, the stub presents an infinite impedance or open circuit at only one frequency and cannot be used for any appreciable bandwidth.

A broad-band stub can be made by changing the impedance of the main line a quarter-wavelength on each side of the stub. (This is illustrated in Fig. 5.4.) In this method of support, the diameter of the center conductor is increased, thus decreasing its characteristic impedance. The length of this low impedance section is a half-wavelength at the center frequency of the band. The supporting stub is attached to the center of

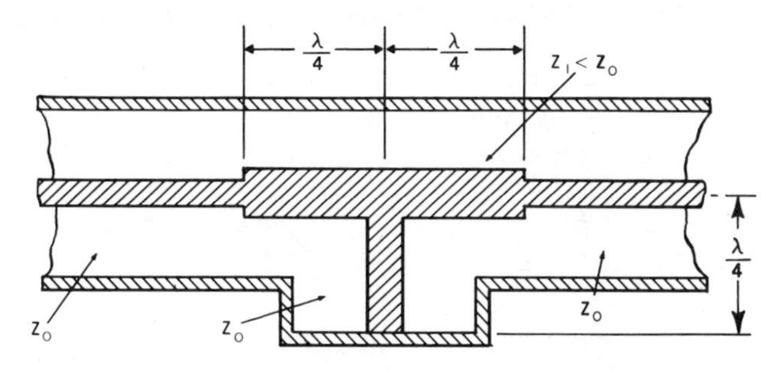

Fig. 5.4. Broad-band stub support.

this low impedance section. At the design center of the band, the stub presents an open circuit, and the half-wavelength low impedance section has no effect on the impedance; that is, as may be seen by letting

$$\beta s = 180°$$

in Eq. (2.33), the input impedance for a half-wavelength line of any finite characteristic impedance is the load impedance.

At a frequency lower than the design frequency, the stub is less than a quarter-wavelength and thus appears as a shunt inductance. This additional inductance effectively adds more length to the low impedance line, which at the lower frequency is *less* than half a wavelength long. At one lower frequency, there will be exact compensation.

At frequencies higher than the design frequency, the stub appears as a shunt capacitance, and the low impedance line is *more* than half a wavelength long. These effects tend to cancel, and again at one higher frequency, there will be exact compensation.

Thus, the broad-band stub is perfectly matched at three frequencies, and the VSWR can be kept low over a broad band by suitable spacing of the three matched frequencies. In practice, stubs have been built with standing-wave ratios less than 1.04 over a 1.5:1 frequency band.

In discussing frequency, it is customary to refer to percentage bandwidths for small frequency ranges. This is obtained by taking the difference between the maximum and minimum frequencies and dividing by the center frequency. Thus, a frequency range from 2775 to 3225 megacycles would be $(3225 - 2775)/3000 = 15$ per cent bandwidth. However, for bandwidths greater than 25 per cent, a more usual method is to divide the maximum by the minimum. Thus, 2000 to 3000 megacycles is a bandwidth of 1.5 to 1. If the maximum is exactly twice the minimum, the range is called one octave.

5.9. BEADS

Another type of support for the inner conductor is a dielectric bead. When a dielectric is the medium between the conductors in a coaxial line, the impedance is lessened. In order to maintain the same impedance,

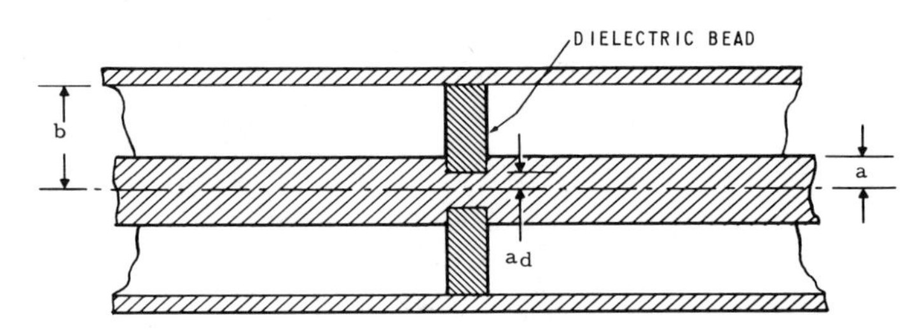

Fig. 5.5. Bead support.

the inner conductor should be cut back as shown in Fig. 5.5. In theory the value of the new radius a_d can be found by equating the impedances.

$$138 \log_{10} \frac{b}{a} = \frac{138}{\sqrt{\epsilon}} \log_{10} \frac{b}{a_d} \tag{5.15}$$

Solving this for a_d,

$$a_d = \frac{b a^{\sqrt{\epsilon}}}{b^{\sqrt{\epsilon}}} \tag{5.16}$$

In practice, however, this does not result in a perfect match because the discontinuities in the inner conductor and the fringing fields produce additional reflections. It has been found empirically that if the new inner conductor radius a_d is decreased about ten per cent, a VSWR under 1.02 will result over a frequency band of many octaves. For example, suppose in an air-filled coaxial line,

$$b = 0.460 \text{ inch}, \qquad a = 0.200 \text{ inch}$$

The characteristic impedance, from Eq. (5.5), is 50 ohms. If a Teflon bead is used to support the inner conductor, the radius of the inner conductor at the bead, a_d, can be found from Eq. (5.16), using $\epsilon = 2.1$ for Teflon. Thus, $a_d = 0.137$ inch. Reducing this by ten per cent to compensate for discontinuities gives the value 0.123 inch for the radius of the inner conductor at the bead. The hole in the bead should be the same size so that the bead supports the inner conductor. The bead may be split in two halves to fit around the smaller diameter, or the rod may be separated at the location of the bead and one end threaded into the other.

In long lines it is necessary to support the inner conductor at more than one point. If many beads are equally spaced, and even if each bead introduces a VSWR of only 1.02, at some frequency the reflections will all be in phase and the resultant VSWR will be 1.02^n, where n is the number of beads. This can be prevented either by randomly spacing the beads or by ascertaining that the high reflection will occur outside the band of interest. The worst reflection occurs when the beads are slightly less than a half-wavelength apart.

5.10. CABLES

Flexible cables are coaxial lines in which a solid, flexible dielectric is used as the support for the inner conductor and as the medium between the two coaxial conductors. The inner conductor is solid, if small, but stranded where greater flexibility is required. It is usually made of copper and is sometimes silver-plated. The outer conductor is made of braided copper and also may be silver-plated. Cables usually have a flexible outer covering to protect the outer braid but this covering has no electrical significance.

Cables have greater attenuation than other coaxial lines. Added losses occur in the dielectric and in the outer braid which has greater resistivity than a solid copper conductor. The attenuation increases with frequency. A typical attenuation value of a "low-loss" microwave cable is 15 decibels per 100 feet at 3000 megacycles.

Most microwave cables have a characteristic impedance in the vicinity of 50 ohms. Bending, flexing, and twisting cause negligible reflections, although severe motions do affect the phase shift (electrical length) through the cable.

5.11. DISCONTINUITIES

When the radius of the inner or the outer conductor of a coaxial line is abruptly altered, the characteristic impedance of the section containing the new dimension is changed. This can be seen directly from Eq. (5.5). It is possible to change both a and b but to keep b/a fixed. In this case, there is no change of impedance. However, in all cases involving changes in conductor size, the steps in the conductors present capacitive discontinuities. When the step is small, the reflection it causes is negligible compared to the effect of the change of impedance.

A thin diaphragm attached to the inner conductor, but not touching the outer conductor is a capacitance. Similarly, a diaphragm connected to the outer conductor, but not touching the inner, is also capacitive. These are illustrated in Fig. 5.6.

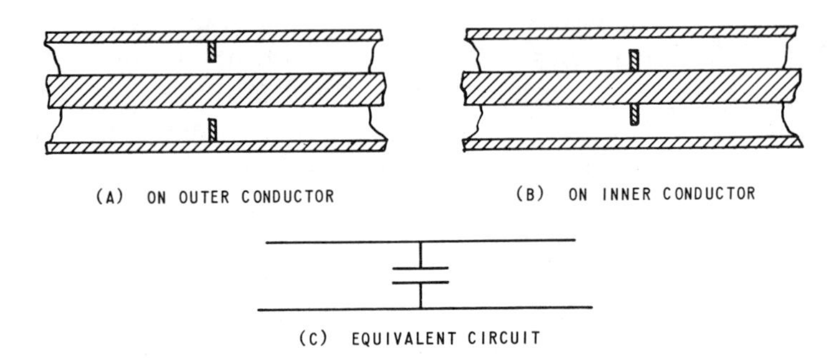

Fig. 5.6. Coaxial diaphragms.

A screw protruding into the coaxial line is a capacitance when first inserted. As the capacitive screw gets closer to the inner conductor, it passes through a series resonance point and becomes inductive. A post connecting the inner to the outer conductor is inductive. However, these two have comparatively small values of reactance since they intercept a small part of the electric field.

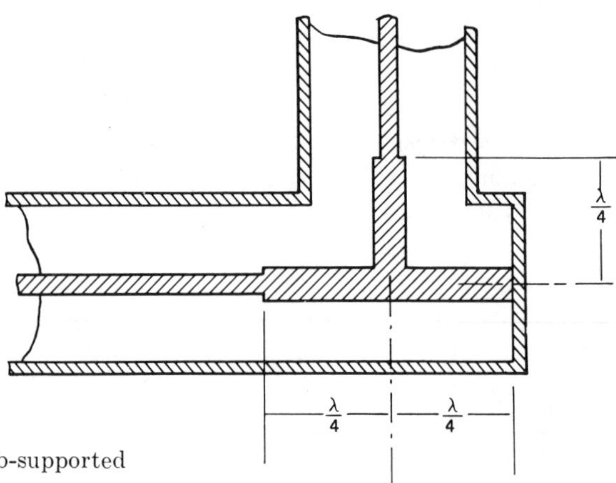

Fig. 5.7. Stub-supported corner.

5.12. CORNERS

A rigid coaxial line can have corners. These can be smooth curves in which the inner and outer conductors are both bent gradually and retain the same cross section around the corner. Such a corner, if well made, should cause negligible reflection at any frequency.

A stub-supported right-angle corner is compact and can be made broadband. It is illustrated in Fig. 5.7; notice that it is very similar to the broadband stub support of Fig. 5.4 and works on the same principle.

5.13. OTHER TEM LINES

Figure 2.2 illustrated some lines which are similar to coaxial lines in that they have two conductors and propagate a TEM mode. When well designed, these "open" coaxial lines have negligible radiation loss. Their major advantages are compactness, ease of fabrication, and the accessibility of the "inner" conductor.

QUESTIONS AND PROBLEMS

5.1. What is the most common mode of transmission in a coaxial line? What is the next higher mode? What is the approximate formula for the cut-off frequency for this mode in an air-filled coaxial line?

5.2. You have a polyethylene ($\epsilon = 2.26$) filled cable. The outer conductor has a diameter of 0.680 in. and the inner a diameter of 0.188 in. Find:
   a. The characteristic inpedance. *ans.* a. 52 ohms
   b. The capacity per foot. *ans.* b. 27.5 pf
   c. Inductance per foot.
   d. The velocity of propagation.
   e. The length of a section a quarter-wavelength long at 100 Mc. *ans.* e. 19.7 in.

5.3. Repeat Prob. 5.2 but assume the line is now air filled.

5.4. Consider the copper coaxial line of Prob. 5.2. The frequency is 3 Gc. The loss tangent for the line at this frequency is 0.00040.
   a. Find the db loss per hundred ft due to conductor losses.
   b. What is the db loss per hundred ft due to the dielectric?
   c. What is the total loss per hundred ft at 3 Gc?

5.5. Explain briefly why 50 ohms is a popular characteristic impedance for coaxial line.

5.6. By means of equivalent circuits show how a broad-band stub may be matched at three frequencies.

5.7. How may multiple reflection effects in a bead supported coaxial line be minimized?

5.8. What is the effect when two cables of the same Z_0 but different diameters are joined together?

5.9. What is the input impedance of a shorted piece of *RG* – (8/*U*) that is six in. long at a frequency of 500 Mc.? What do you consider to be important sources of error in your calculations for this problem?

6

METHODS OF MATCHING

Any mismatch in a line causes a reflection of some of the incident energy which represents a decrease in the amount of power reaching the load. The standing-wave ratio is an indication of the magnitude of the reflection, and, consequently, one of the important design problems is to reduce the VSWR to as near unity as practicable. There are, in general, two methods of matching. In one, the mismatched load impedance is transformed to be equal to the source impedance so that the source sees a matched load. In the second method, a discontinuity is placed in the line with a reflection equal to (in amplitude) and out of phase with the reflection already existing in the line. The two reflections thus cancel, and all power goes to the load.

In practice, impedance as close to the load as possible is plotted on a Smith Chart as was explained in Sec. 3.7. Then matching devices are used to bring the impedance into the center of the chart.

6.1. QUARTER-WAVE TRANSFORMERS

Equation (2.37) indicates that a quarter-wavelength section of line can be used to transform a real impedance to any other real impedance. The characteristic impedance of the transformer section, Z_0, is determined by Eq. (2.37) as follows:

$$Z_0^2 = Z_s Z_L \tag{6.1}$$

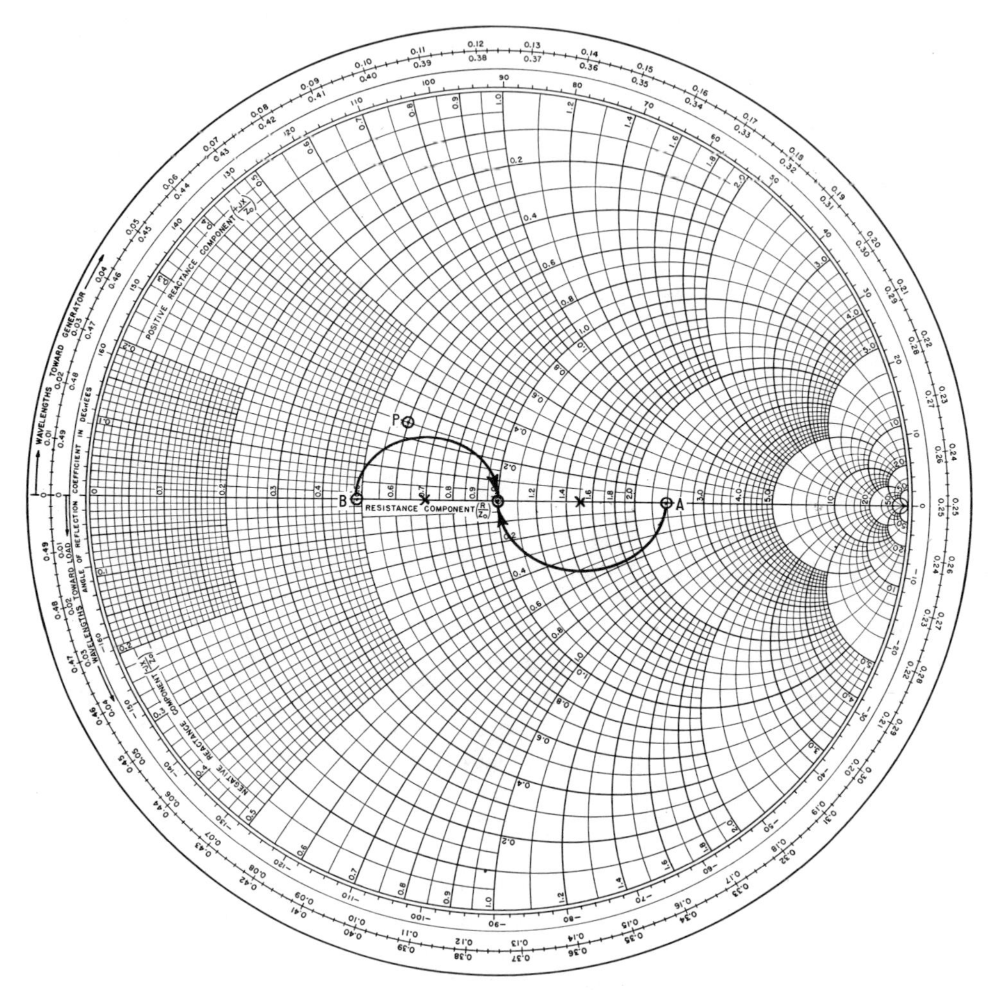

Fig. 6.1. Quarter-wave transformer.

When this is normalized to the input impedance it becomes

$$z_0^2 = z_L \tag{6.2}$$

since z_s is unity. (It is assumed that Z_s is the source or generator impedance to which other impedances are referred.)

Thus, if the impedance plotted on the Smith Chart is real, that is, if it lies on the horizontal diameter of Fig. 6.1, it can be brought to the center by a quarter-wave transformer. Point A at $R = 2.4$ in the figure represents a real impedance higher than the source impedance. The impedance can be brought into the center by a quarter-wave transformer with characteristic impedance equal to 1.55 ($= \sqrt{2.4}$). This is indicated by a cross on the chart. Since the impedance represented by point A is close to the load, the quarter-wave transformer begins at the point in the transmission line corresponding to point A in the chart and progresses toward the generator. In effect point A rotates, as shown by the arrow, around a new impedance, represented by the cross at $R = 1.55$. The direction of rotation is clockwise since this represents movement toward the generator. It should be remembered that one revolution on the Smith Chart represents a half-wave movement in the line, so that a quarter-wave is half a revolution.

Point B at $R = 0.49$ in Fig. 6.1 represents a real load impedance lower than the source impedance. Like point A, this is rotated around the normalized impedance of 0.7 ($= \sqrt{0.49}$), this also is represented by a cross.

This method of matching is useful at a spot frequency or for a very narrow frequency band. Obviously, if the frequency spread is large, the matching section, which is a quarter-wavelength at mid-band, will be less than a quarter-wavelength at the low frequency limit and more than this at the high end. (This is shown in Fig. 6.2.) Point A again represents a load impedance, this time at $R = 4$. A quarter-wave transformer with characteristic impedance equal to two will indeed bring the mid-frequency into the center of the chart, but the high frequency end will travel more than a quarter-wavelength and might end at a point represented by H, while the low end stopped at L. The line from L to H would then represent the impedance over the band. The band edges have a VSWR of approximately 1.3.

If 1.3 is above specifications, the situation can be improved by the use of two quarter wave transformers as shown in Fig. 6.3. The first transformer is chosen at some arbitrary value such as $R = 3$. This means the transformer's characteristic impedance is three times the source impedance. The rotation will then be to $R = 2.25$ for mid-band ($4 \times 2.25 = 3^2$). But the high and low frequency ends will travel more and less than a quarter-wavelength, respectively, and end at H′ and L′ as indicated. A second transformer is now used to bring the impedance to the center. Since the load now looks like 2.25 for mid-band, this second transformer has a characteristic impedance of 1.5 ($= \sqrt{2.25}$). Now mid-band will rotate right to the center of the chart. The high end will travel more than a quarter-wavelength but it has more than a quarter-wavelength to go to reach the center. Thus, the frequency errors of the two transformers

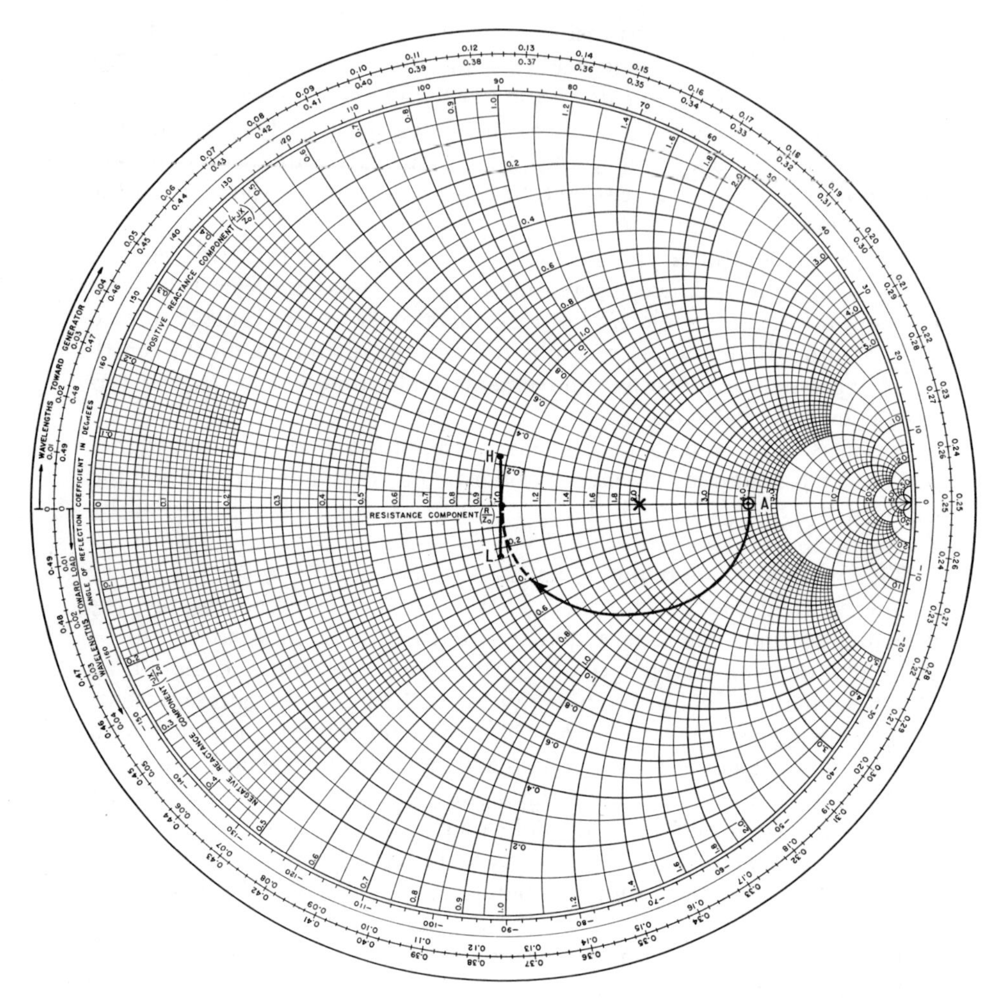

Fig. 6.2. Frequency sensitivity of $\frac{\lambda}{4}$ transformer.

tend to compensate, and the result is a better match. The equivalent circuit is shown in the upper left of Fig. 6.3.

It can be shown that the greater the number of quarter-wave steps, the wider the frequency range for a given maximum VSWR. The design of multistepped transformers consists of calculating the minimum number of transformers and finding characteristic impedance of each in order to achieve the specified VSWR in the frequency band. To describe the entire designing process would be beyond the scope of this book. However,

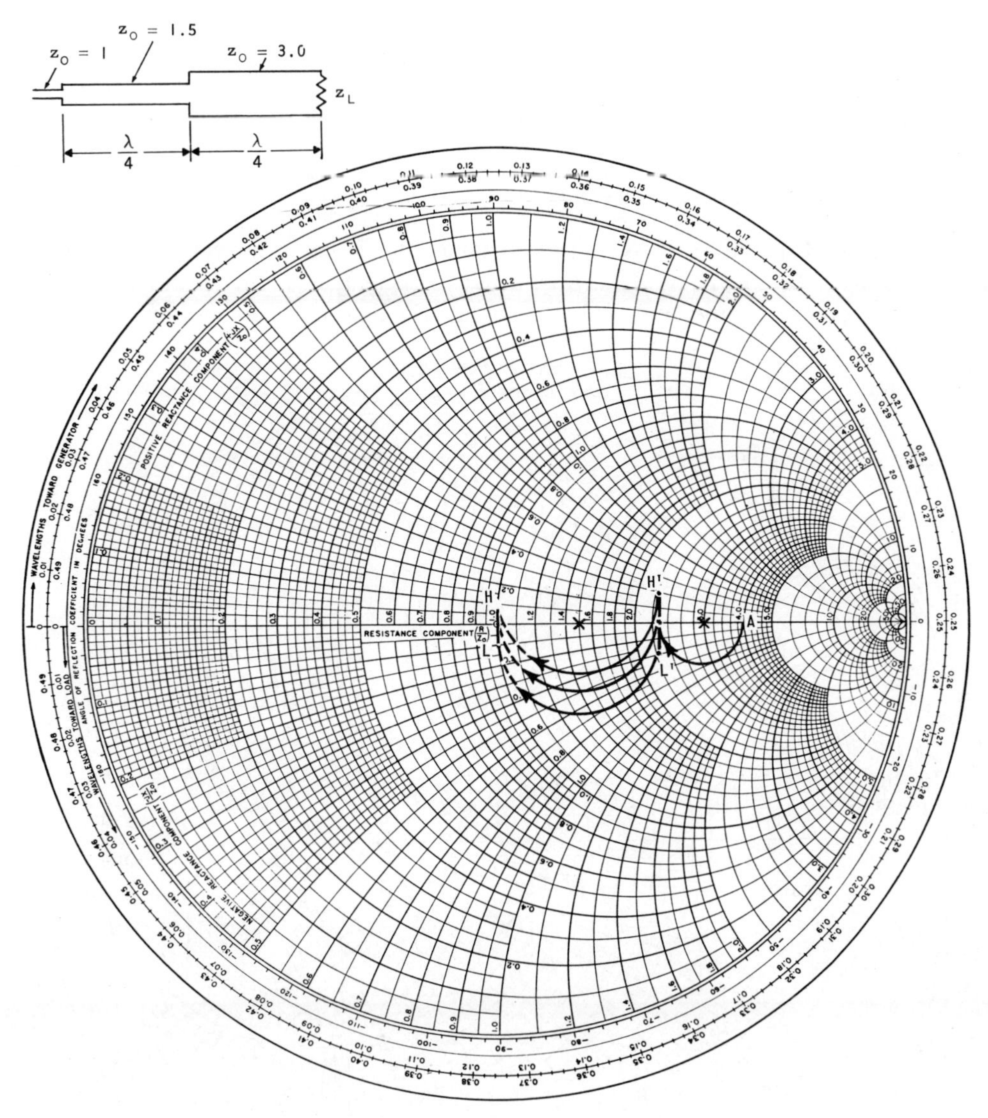

Fig. 6.3. Double $\frac{\lambda}{4}$ transformer.

a two-section transformer having the broadest bandwidth can be designed from the following equations:

$$z_1 = (z_L r)^{1/4}$$
$$z_2 = \frac{z_L}{z_1} \tag{6.3}$$

where z_1 is the characteristic impedance of the $\lambda/4$ step nearer the source, z_2 is that of the step nearer the load, z_L is the load impedance, and r is the specified maximum VSWR. All impedances are normalized to the source impedance.

6.2. HALF-WAVE TRANSFORMERS

In Eq. (2.33), if $\beta s = 180°$ (that is, if s is half a wavelength), then $\tan \beta s = 0$, and $Z_s = Z_L$. Thus, a half-wavelength section of line of any characteristic impedance will not change the load impedance. For a broad

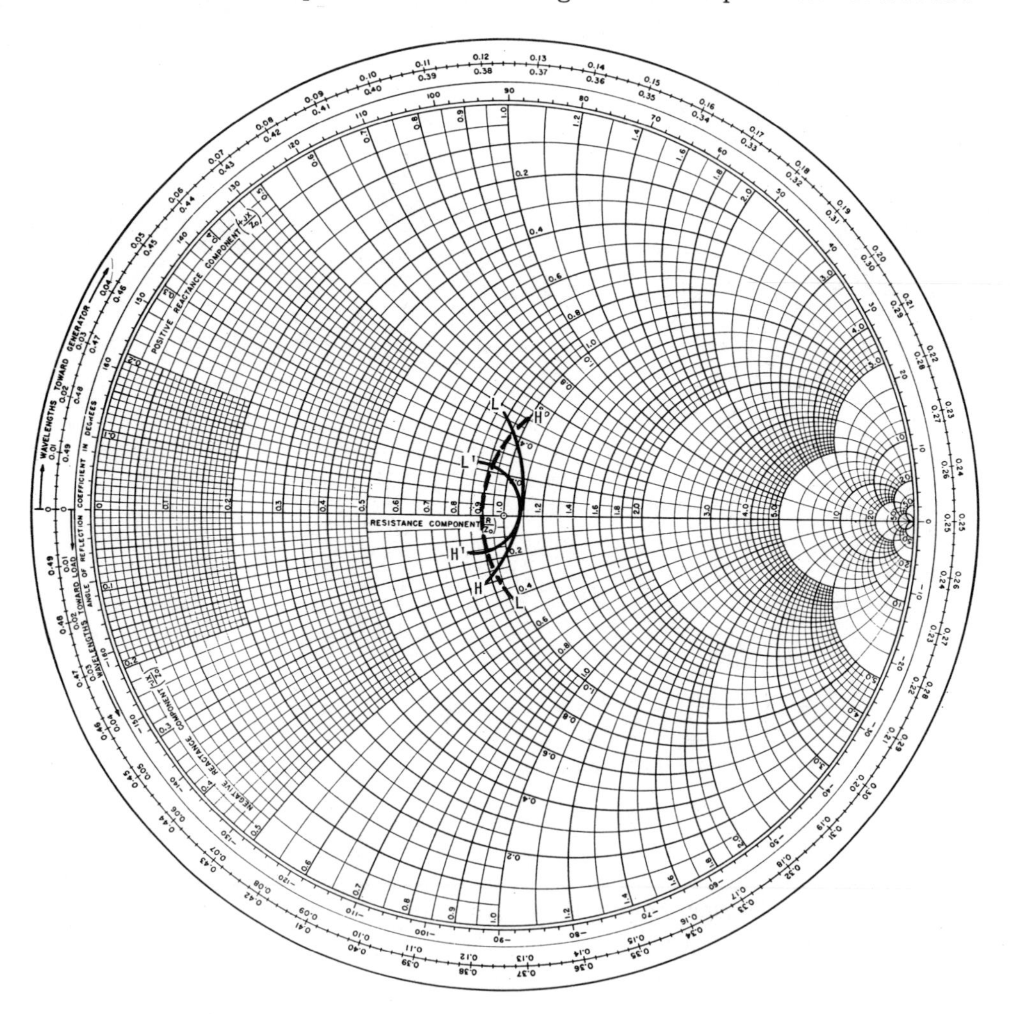

Fig. 6.4. Half-wave matching.

band of frequencies, however, a given length of line can be exactly half a wavelength long at only one frequency and will produce impedance changes at all other frequencies. A half-wave transformer is useful when the center frequency is already matched, and the high frequency end can be matched by a rotation of more than half a wavelength while the low end requires less. (The high end always travels farther than the low end because it has a shorter wavelength.) Figure 6.4 shows two possibilities.

The solid line from H to L in Fig. 6.4 represents the impedance of a load as a function of frequency. The high end of the band is capacitive, and the low is inductive but the mid-band frequency is close enough to center. If this is rotated half a wavelength around some point of higher characteristic impedance, such as 3, the center will stay fixed, but the high end will move more than a half wave to H′, for instance, while the low end moves to L′. The resultant impedance plot has a lower VSWR over the band.

The dotted line in Fig. 6.4 represents the case when H is inductive and L is capacitive. By reasoning similar to the above, the half-wave transformer must have a lower characteristic impedance than the source in order to close up the impedance plot.

6.3. TAPERS

It is possible to change from one impedance to another by using a continuous taper instead of transformers. The taper may be a simple linear transformation or it may be shaped so that a plot of characteristic impedance as a function of position along the taper is an exponential curve, a hyperbola, or some other special function. In practice, shaped tapers are difficult to manufacture and are seldom used. Although special shapes can result in a good match in a short length, for tapers longer than approximately two wavelengths there is little difference in VSWR between any of the special shapes.

Quarter-wave transformers will usually result in a better VSWR in a given length than any taper. Thus two $\lambda/4$ transformers will have a better VSWR than a half-wavelength taper. But in the laboratory, where space is not at a premium, long tapers are usually used in preference to transformers because of the relative ease of fabrication.

6.4. SHORT TRANSFORMERS

It is unusual for the load impedance to be a pure real number. More frequently, it will have a complex value such as that shown by point P in Fig. 6.1. Since this point lies near the arc drawn around $R = 0.7$, it is reasonable to suppose that a transformer with this value of normalized characteristic impedance should cause point P to move somewhere close to the center. The length of the transformer must be less than a quarter-wavelength, since it is required to rotate point P only as far as the real axis.

It should be noted that rotations about the center of the Smith Chart are circles; that is, if the center of the chart corresponds to the characteristic impedance of the transmission line, the locus of the impedance seen at any point in the line will be a circle with its center at the center of the chart. It is customary to speak of rotations around other points when using transformers; however, *these* rotations are elliptical. To find their exact paths it is first necessary to move the characteristic impedance of the transformer to the center, describe the circle, and then return the characteristic impedance to the point where it belongs. This can best be demonstrated by an example.

In Fig. 6.5, the point P of Fig. 6.1 is repeated at $0.6 + j0.25$. The first approximation, made by eye, shows that this point can be rotated about 0.7 to land at the center. If the transformer has a characteristic impedance which is 0.7 of the source impedance, then the chart should have this value at the center during the transformation. Thus, if the source impedance is 50 ohms, the new value of the center is 35 ohms. The point P represents the impedance referred to or normalized to 50 ohms. The new value, referred to 35 ohms, must be found from the relationship

$$50(0.6 + j0.25) = 35(R + jX)$$

Thus

$$R + jX = 0.86 + j0.36$$

and is shown on Fig. 6.5 as point Q. It should be clear that the impedance represented by point Q is normalized to 35 ohms and is thus the same as the impedance of point P, which is normalized to 50 ohms. For point Q, the center represents 35 ohms. The same results would be achieved without using actual ohmic values by using the normalized relationship

$$1(0.6 + j0.25) = 0.7(R + jX)$$

Now point Q is rotated around the center until it is on the real axis at point S, corresponding to a value of $R = 1.5$. The amount of rotation can be found by first drawing a line from the center through point Q. The line intersects the outside circle at 0.114; the real axis through S intersects the outside circle at 0.250. The total rotation is the difference, 0.136λ.

Point S is still referred to 0.7 of the source impedance and now must be returned to 1.0. The relationship is reversed. $0.7(1.5) = 1.0(R)$. Thus R is 1.05, and S referred to 0.7 is now changed to point T which is normalized to the original source impedance.

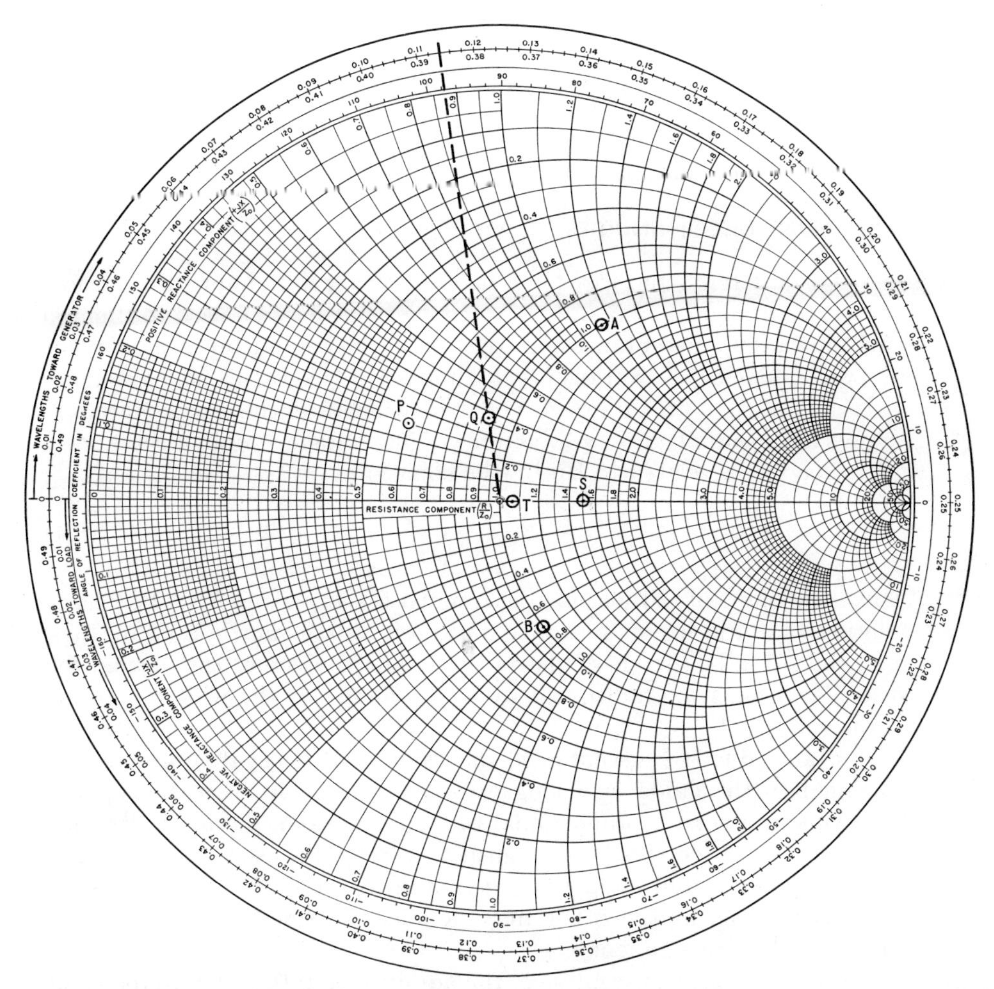

Fig. 6.5. Short transformer.

The transformer calculated above has a length of 0.136λ and a characteristic impedance which is 0.7 of the source impedance. It is not perfect because it did not bring point P directly to the center. However, if the transformer's impedance were slightly less, point T would be nearer the center. It would be unnecessary to repeat the calculations; a transformer would be built with a characteristic impedance of 66 or 67 ohms.

The procedure above was illustrated for one reason. If a band of frequencies must be matched, it is usually sufficient to plot the two frequency limits and the center frequency and to perform the calculations on these three points. The length of the transformer is usually determined to match the middle frequency. At the band edges, then, the transformer will have different electrical lengths, and these should be taken into consideration. When one transformer will not match the whole frequency band, this can usually be done by a combination of two or more short transformers.

6.5. SHUNT DISCONTINUITIES

Posts, screws, and irises in a waveguide act like shunt susceptances (as indicated in Figs. 4.5, 4.6, 4.7, and 4.8) and may be used to "match out" unwanted susceptances. If the normalized admittance at a point in a waveguide or other transmission line is $1 \pm jB$, then a shunt element at this point with an admittance of $\mp jB$ will cancel the susceptance and the resultant admittance will be unity. The voltage standing-wave ratio of the load, r, is related to $|B|$ by the equation

$$|B| = \frac{r - 1}{\sqrt{r}} \tag{6.4}$$

Thus, the value of $|B|$ may be determined for any given VSWR. Charts relating B to iris size or post size and position are shown later in Figs. 6.10 and 6.11. To use these charts the required value of the normalized susceptance must be multiplied by a/λ_g, where a is the wide dimension of the waveguide.

For an inductive susceptance, the value of B is negative and is proportional to the guide wavelength. For a capacitive susceptance, it is positive and inversely proportional to λ_g. Thus, shunt discontinuities are frequency sensitive, but their sensitivity may be used to advantage. Since capacitive and inductive susceptances have opposite frequency sensitivities, one or the other may be used to cancel out a mismatch which varies with frequency.

In order to use a susceptance, the admittance rather than impedance is plotted. The admittance of the load must be rotated until its plot on the Smith Chart lies on the susceptance circle which passes through the center. In Fig. 6.5, points A and B are admittances. The admittance of point A is $1.0 + j1.14$; that of point B is $1.0 - j0.68$. An inductive post or iris with susceptance equal to $-j1.14$ will match the admittance of point A. A capacitive iris or tuning screw with susceptance equal to $+j0.68$ will match that of point B.

For broad-band matching, the exact method depends on shape of the admittance plot. Figure 6.6 shows some typical cases. It must be remembered that the plot is usually made close to the load so that matching has to be done by moving toward the generator or in a clockwise direction. In moving along the line, the high frequency end moves faster than

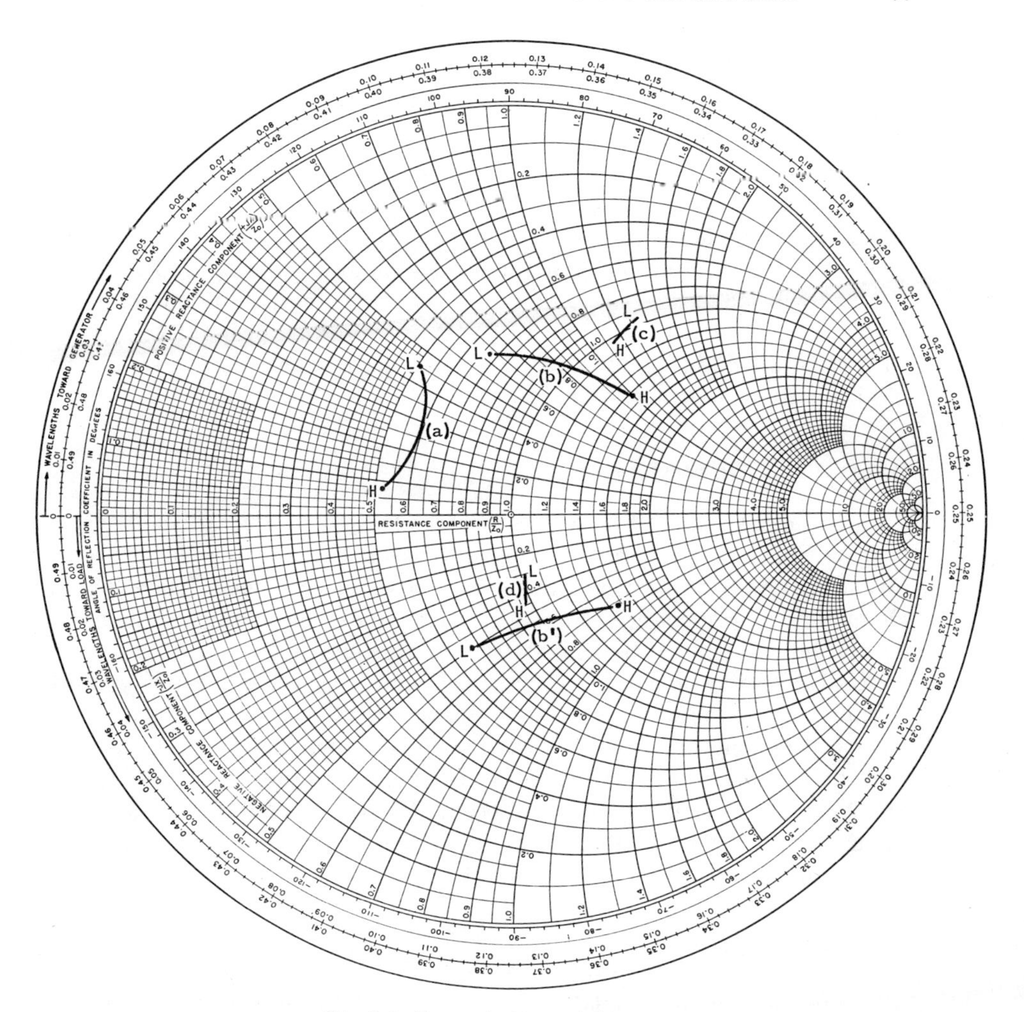

Fig. 6.6. Susceptance matching.

the low end. Curve (a) in Fig. 6.6, shows that low end is "ahead" of the high. As the susceptance rotates in a clockwise direction, the high end will tend to overtake the low end, and the curve will be decreased. The curve is rotated until it is as closely packed as possible on the circle passing through the center. Then the proper value of susceptance will move it into the center.

When the high end is ahead of the low, rotation toward the generator will spread the curve rather than close it. To reverse the curve, it is first moved to a place where it straddles the $R = 1.0$ line, as indicated by curve (b) in Fig. 6.6. Now a large inductive susceptance will move the whole curve down to the lower half of the Smith Chart, as indicated by curve (b′). Further rotation will close up the curve, since the low frequency end now leads.

If the high frequency end has the lower VSWR, ideally, the admittance curve should be brought to a position on the upper side of the $R = 1.0$ circle, as indicated in curve (c). Now an inductive susceptance will close the curve further, since its B varies with λ_g, and thus the low end will move into the center faster than the high. Obviously, if the high frequency end has a higher VSWR, it should be brought to a point such as is indicated in curve (d), and a capacitive susceptance should be used to correct the mismatch.

6.6. CHOICE OF MATCHING

In coaxial lines it is comparatively simple to change impedance. A short sleeve on the inner conductor serves as a lower impedance transformer; a new section of smaller diameter inner conductor has a higher characteristic impedance. For this reason most matching problems in coaxial line are solved by impedance transformers.

In waveguides, changing impedance is more involved, requiring more complicated machining, and sometimes introducing problems of new cut-off wavelengths. For this reason susceptance matching is usually preferred in waveguide problems. Inductive posts and capacitive screws are the simplest and are thus preferred. However, for high power work, capacitive matching elements should be avoided, since they lower the breakdown power of the waveguide.

6.7. EXAMPLES

For convenience in the following examples, the VSWR is always less than two to one. On a standard Smith Chart, the plots would be within a very small circle near the center, and would consequently be difficult to manage. For this reason an expanded chart, which is simply the center portion of a much larger chart, is used.

Example 6.1. The impedance plot of a coaxial line component from 3.0 to 3.3 kilomegacycles is shown in curve A of Fig. 6.7. The ends of the curve are labeled with the frequencies, although they could just as conveniently be labeled H and L for high and low. The center frequency, 3.15 kilomegacycles, is also indicated. The reference is 50 ohms, since measurements were made with a 50-ohm slotted line. Since this is a plot of impedance close to the component, the matching devices must be put in toward the generator, which means that the rotation on the chart will be clockwise. At first glance, a transformer

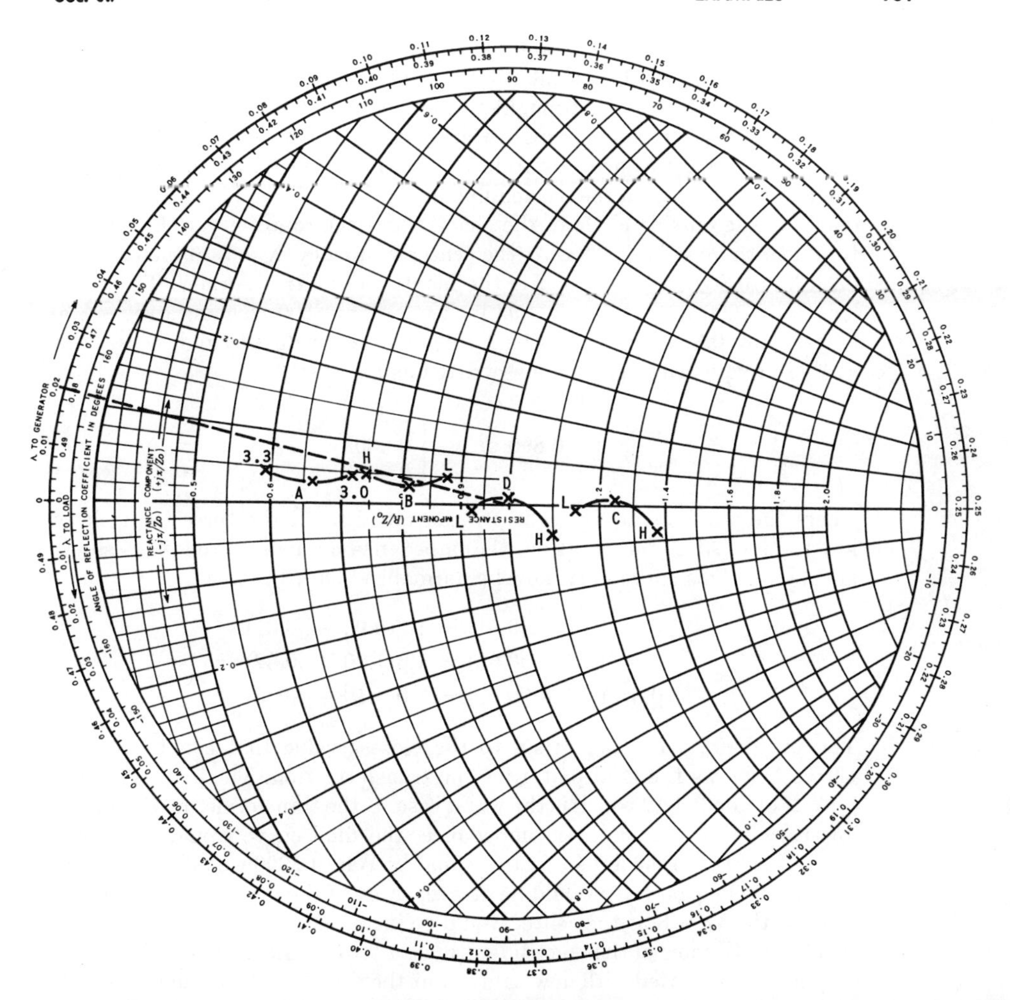

Fig. 6.7. *Example* 6.1.

slightly less than a quarter-wavelength at the center frequency with a characteristic impedance about 0.81 (normalized) should bring this close to the center. (The center frequency impedance is $0.66 + j0.03$. The value 0.81 was obtained by taking the square root of 0.66.) Using this method, the first step is to normalize the plot to 0.81 instead of unity. The values of impedance for the three frequencies, low to high, are written and normalized.

$$0.72 + j0.04 = 0.81(0.88 + j0.05)$$
$$0.66 + j0.03 = 0.81(0.81 + j0.04)$$
$$0.59 + j0.04 = 0.81(0.74 + j0.05)$$

The new curve, normalized to 0.81, is curve B in Fig. 6.7. This curve must now be rotated to the right to the real axis, the horizontal diameter. If the center frequency is brought to the real axis, the high end will pass it, whereas if the high end is brought there, the rest of the curve won't reach it. Some compromise value of rotation is chosen arbitrarily. In this case, it was decided to rotate from the dashed line in the figure. This line passes through 0.02λ on the outside circle of the chart. Since the rotation is to 0.250λ, the total length must be 0.230λ. At the center frequency, $\lambda = 3.74$ inches. Thus $0.230\lambda = 0.860$ inches, which is the length of the transformer. The wavelength at 3000 megacycles is 3.93 inches, and at 3300 megacycles it is 3.57 inches. Now 0.860 inches must be divided by these values to find the rotation at these frequencies. This is now written as follows:

$$0.860 \text{ in.} = \begin{cases} 0.219\lambda \\ 0.230\lambda \\ 0.246\lambda \end{cases}$$

where the numbers at the right are the rotations at the three frequencies, low to high. After this rotation, the impedance normalized to 0.81 is shown in curve C. This curve must now be normalized back to unity.

$$\begin{aligned} 0.81(1.14 - j0.01) &= 1.0(0.93 - j0.01) \\ 0.81(1.23 + j0.02) &= 1.0(1.00 + j0.02) \\ 0.81(1.36 - j0.07) &= 1.0(1.10 - j0.06) \end{aligned}$$

The imaginary parts are taken to the closest value that can be plotted. Curve D shows the final plot after the transformation. The high end has a VSWR of about 1.12 which is the worst case. If the transformer had a slightly lower value of characteristic impedance, the final curve would have been moved slightly to the left so that the final VSWR at 3300 would be less while that at 3000 megacycles would be greater. Also, if the transformer is slightly shorter, the VSWR at the center would be increased, but at 3300 it would be decreased. In view of these considerations, after the calculations are finished, they are not repeated with new values, but the transformer is usually made to slightly different values. In this case, the transformer would be built with a characteristic impedance of 40 ohms (0.80 instead of 0.81) and a length of perhaps 0.850 inches.

Example 6.2. If the VSWR limit is 1.10, then the matching scheme used in Ex. 6.1 will just barely meet specifications with no margin for error. Figure 6.8 shows a more complicated matching scheme which will result in a lower VSWR. Curve A in Fig. 6.8 is the same plot as curve A in Fig. 6.7. First the curve is rotated at 50 ohms, without any change, to a more convenient portion of the chart. This can best be explained by illustration. Let the first rotation be 0.250 inches long. The three wavelengths, as before are 3.93, 3.74, and 3.57 inches. Thus,

$$0.250 \text{ in.} = \begin{cases} 0.064\lambda \\ 0.067\lambda \\ 0.070\lambda \end{cases}$$

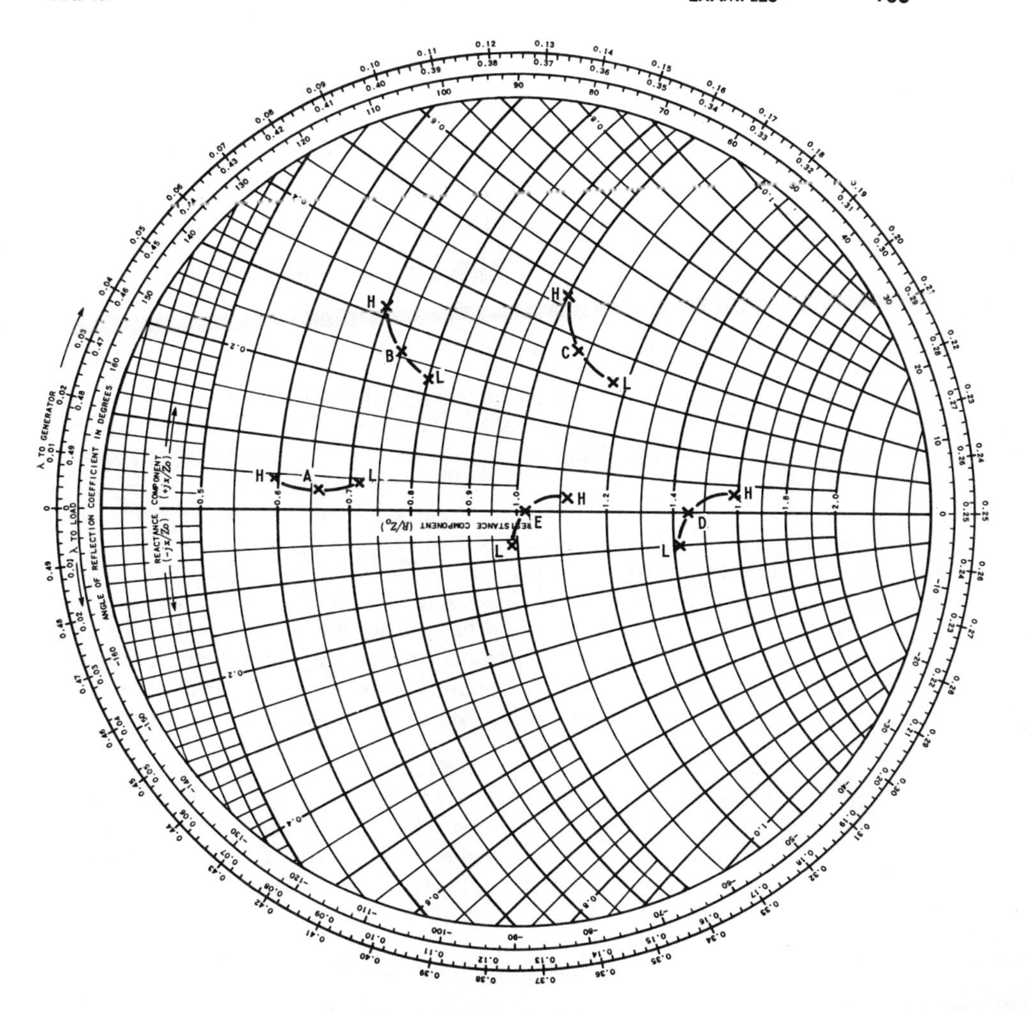

Fig. 6.8. *Example* 6.2.

After this rotation the plot is at curve B in Fig. 6.8. Now if the curve is rotated around approximately 0.7, it should settle close to the middle and the high end should overtake the lower because it moves faster. First, curve B must be normalized to 0.7.

$$0.81 + j0.23 = 0.7(1.16 + j0.33)$$
$$0.75 + j0.26 = 0.7(1.07 + j0.37)$$
$$0.70 + j0.32 = 0.7(1.00 + j0.46)$$

This is plotted as curve C and rotated until it is on the real axis, approximately 0.365 inches.

$$0.365 \text{ in.} = \begin{cases} 0.093\lambda \\ 0.098\lambda \\ 0.102\lambda \end{cases}$$

This last is, of course, normalized to 0.7; it must now be returned to unity.

$$\begin{aligned} 0.7(1.41 - j0.10) &= 1.0(0.99 - j0.07) \\ 0.7(1.44 + j0) &= 1.0(1.01 + j0) \\ 0.7(1.59 + j0.05) &= 1.0(1.11 + j0.04) \end{aligned}$$

Curve E shows the final result, achieved by using a transformer which has a characteristic impedance of 35 ohms (0.7) and is 0.365 inches long. The transformer was spaced 0.250 inches from the point where the original match was measured.

Curve E in Fig. 6.8 is not appreciably better than curve D in Fig. 6.7, but the method of Fig. 6.8 can be used to improve the match still further. If the rotation from curve A to curve B in Fig. 6.8 had been normalized to a higher value, say 60 or 65 ohms, instead of just rotating at 50 ohms, curve C would be more nearly horizontal so that the low end and the high end would be equidistant from the center. Then the final VSWR would be under 1.05. In practice, after a preliminary calculation as above, a new calculation is performed with the first rotation around 65 ohms instead of 50 ohms.

Example 6.3. In 0.900 by 0.400 inch waveguide (X-band), the admittance of a component referred to the waveguide admittance is plotted as curve A in Fig. 6.9. The frequency band is 8.5 to 9.5 kilomegacycles. Since this is waveguide, the wavelength used must be guide wavelength rather than free-space wavelength. The guide wavelengths for the three frequencies can be found from Eq. (4.2) or from convenient tables. They are

$$\begin{aligned} \text{at 8500 Mc, } \lambda_g &= 2.18 \text{ in.} \\ \text{at 9000 Mc, } \lambda_g &= 1.91 \text{ in.} \\ \text{at 9500 Mc, } \lambda_g &= 1.72 \text{ in.} \end{aligned}$$

Presumably, this could be matched by using a quarter-wave matching section with a normalized admittance of 1.35. (A quarter-wavelength in the waveguide at 9000 megacycles is 0.472 inches.) This can be done by putting a block in the waveguide so that the height is constricted to $0.400/1.35 = 0.296$ inch. The width, 0.900 inch remains the same so that there is no change in the guide wavelength. The difficulty with this method is that the face of the block adds a capacitive susceptance which introduces too great an error.

In working with waveguide it is better to match *entirely* by using susceptances. To do this curve A must first be moved to the R = 1.0 line. The center frequency is rotated to this line first to determine the distance to move. In this case it is from 0.248 on the outside circle to 0.352, or a total

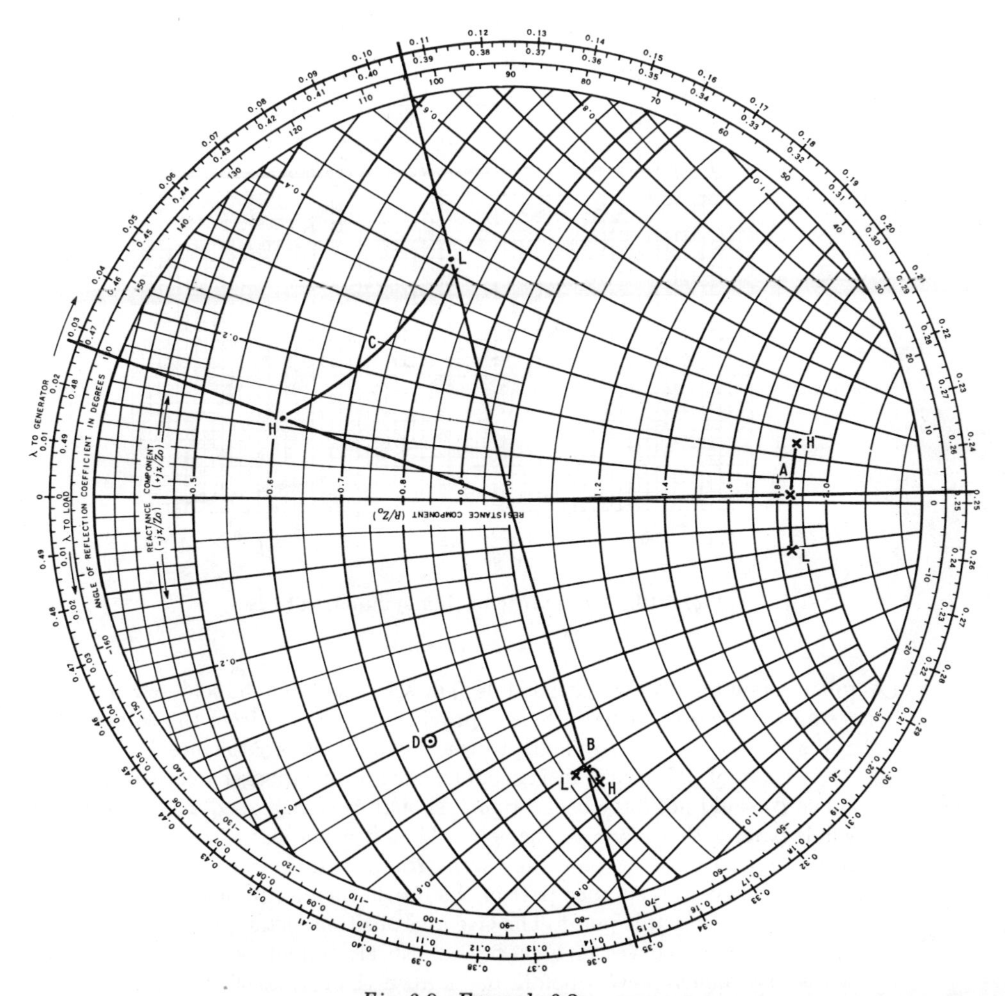

Fig. 6.9. *Example* 6.3.

rotation of $0.104\lambda_g$, 0.199 inch ($= 0.104 \times 1.91$). The other points can be rotated the same distance.

$$0.199 \text{ in.} = \begin{cases} 0.091\lambda_g \\ 0.104\lambda_g \\ 0.116\lambda_g \end{cases}$$

The rotated plot is curve B in Fig. 6.9. It should be noticed that the plot has closed up because the high end travels faster than the low. A capacitive susceptance at this point should move the curve right into the center. The simplest susceptance to use here is a small screw such as a No. 4-40 or a No. 6-32. This is screwed into the waveguide, while the VSWR is observed,

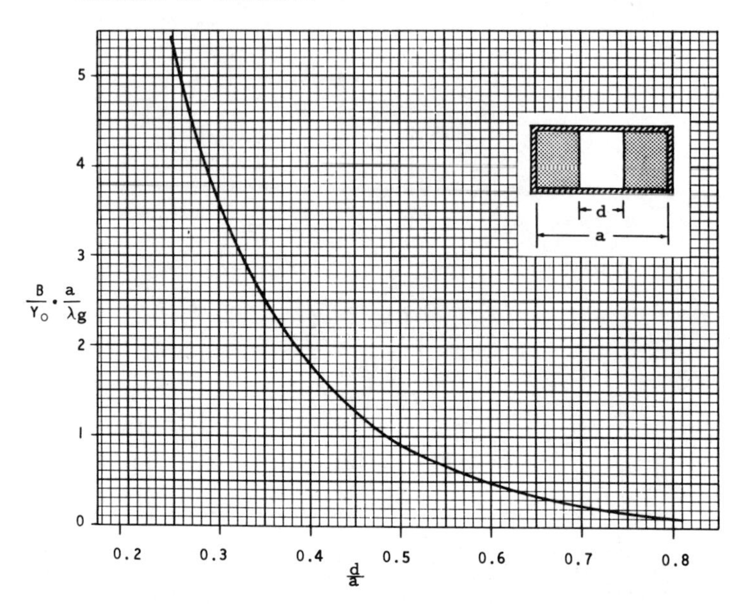

Fig. 6.10. Susceptance of thin symmetrical irises.

until the match is as low as possible. The screw can then be soldered in place. Using this method, a screw hole is provided 0.199 inch from the component and a screw is pushed to the depth where the match is satisfactory.

If the plot of curve A in Fig. 6.9 covered a much larger frequency band, curve B would not have been "tight" enough to give a good match over the whole band with a screw at this point. Then it would have been necessary to rotate farther until the plot was a tight group. If this occurred near $R = 1.0$ in the positive susceptance part of the chart, then an inductive post or iris would be used at this point. The size and position of the post can be determined from the curves in Figs. 6.10 and 6.11. Inductive susceptances will withstand higher powers before breakdown than capacitive and are therefore preferred whenever power handling capability is a consideration.

It is a simple matter to determine how far the admittance must be rotated in order to have the high frequency end overtake the low. Let d be the distance in inches of the rotation. The high frequency end will rotate d/λ_{gH} on the chart where λ_{gH} is the guide wavelength at the high frequency. Similarly, the low frequency end will rotate d/λ_{gL}. The difference between the two is the excess amount that the high frequency has to travel to overtake the low. For example, in Fig. 6.9, curve C is an admittance plot from 8.5 to 9.5 kilomegacycles. A line passes through 0.106λ on the outside circle and the corresponding line for the high end passes through 0.027λ. Thus, the high end must travel 0.079 wavelength ($0.106 - 0.027$) more than the low to land at the same angle from the center. Thus,

$$\frac{d}{\lambda_{gH}} - \frac{d}{\lambda_{gL}} = \frac{d}{1.72} - \frac{d}{2.18} = 0.079$$

and d = 0.643 inch. The spot at which this closing takes place is indicated by D on the chart. In practice, since this is nearer the negative susceptance part of the R = 1.0 line, a rotation slightly less would be used in order to stop right on the line. Then a capacitive susceptance would be used to match the line.

6.8. EMPIRICAL MATCHING

When the mismatch on a line is not too great, it is sometimes possiblc to match the line by using a cut-and-try approach. However, since any change which improves the match at one frequency might make it worse at another, it is necessary to watch a presentation of the match across the whole frequency band while doing this. This can be done by using the sweep oscillator set-up of Fig. 3.10. As a tuning screw is adjusted, the change in VSWR across the whole frequency band can be observed on the oscilloscope. Effects of inserting a post in a line or moving a

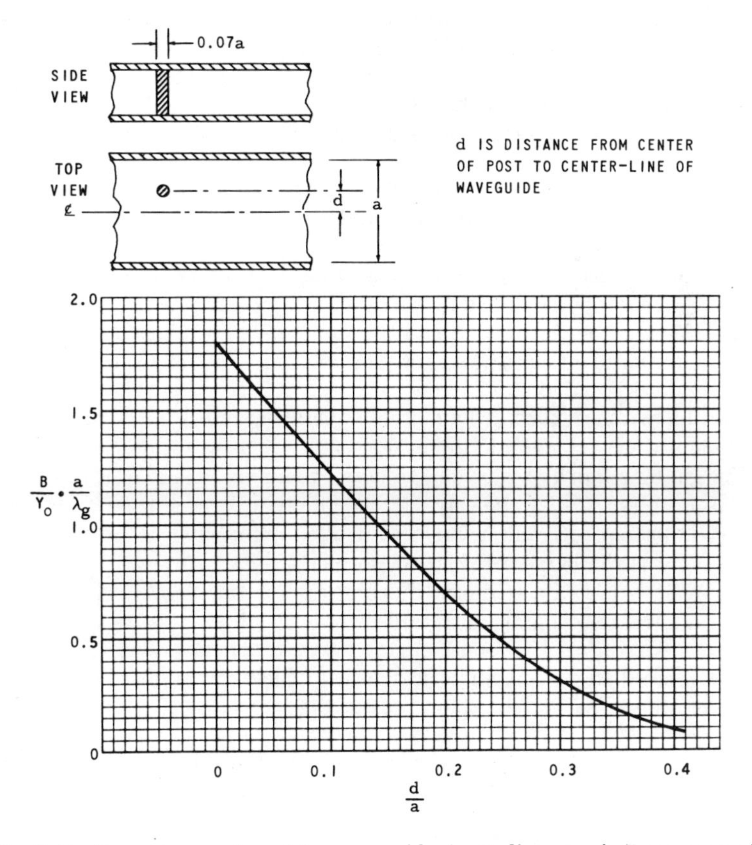

Fig. 6.11. Susceptance of post in waveguide (post diameter is 7 per cent of width of guide).

discontinuity are visible. By noting the individual effects, it is frequently possible to match a component or network, using two or more discontinuities, without the necessity of time-consuming plotting and calculations.

QUESTIONS AND PROBLEMS

6.1. What are the two general methods of matching? Discuss their similarities.

6.2. Assume you have a load of 150 ohms that is to be matched to a 50-ohm line by means of a quarter-wave transformer. The center frequency is 500 Mc.
a. What is the proper surge impedance for the transformer?
ans. a. 86.6 ohms
b. What is the length of the coaxial transformer if the dielectric constant of the line used is 2.25? *ans.* b. 3.95 in.

6.3. In Prob. 6.2 assume that the bandwidth is now to be 30 per cent.
a. Make a plot on a Smith Chart of the impedance (normalized) the input to the transformer presents to the 50-ohm line. Label the end points and center frequencies on this plot.
b. What is the VSWR at the endpoint frequencies?

6.4. Design a transformer to match $30 - j20$ ohms to a 50-ohm line.
a. What is the surge impedance and length of this transformer?

6.5. The normalized load as seen by a guide is $1.5 + j0.6$ ohms. It is to be matched by a susceptance in the guide.
a. How far from the load would the susceptance be placed. *ans.* a. 0.2λ
b. What value of susceptance would be used? *ans.* b. $-j0.65$ mhos

6.6. You have an air-filled X-band guide operating in the $TE_{1,0}$ mode. The frequency is 10 Gc. (Guide dimensions 0.9×0.4 in. inside.) The VSWR in the guide is 2.5. It is decided to correct this by means of a quarter-wave slab of loss-free solid dielectric in the guide.
a. Find the λ_g, λ_0, and λ_c for the air-filled main guide.
b. What is the length of the transformer?
c. What is the proper dielectric constant for the slab? Can you find any material that might be suitable for this?

6.7. When is a half-wave transformer of use? Show an application on a Smith Chart plot.

6.8. There is an obstruction in a guide. Explain how you could determine the nature of the obstruction by VSWR measurements.

6.9. Calculate the β per cm for the main waveguide and transformer of Prob. 6.6.

6.10. A load has the normalized value $0.6 + j0.4$. The guide is to be made flat by means of a transformer. What is the length and normalized impedance of the transformer? Could this be done by means of a dielectric? Explain.

ans. $0.067\lambda_g$, $z = 0.45$ ohms

6.11. A load of $160 - j80$ ohms is to be matched by means of a susceptance to a 100-ohm line. What susceptance is required and where should it be placed?

ans. $-j80$ mhos, 0.105λ from load

6.12. A load has the following values:

At frequency	Impedance
8.4 Gc	$0.5 + j0.4$
10.0 Gc	$0.44 + j0.3$
12.0 Gc	$0.4 + j0.2$

a. What is the VSWR at each frequency?
b. Where would a shunt susceptance be placed to minimize the VSWR over the band? What is the value of the required susceptance?
c. Give the details for a transformer.
d. Discuss the results of *b* and *c*. Point out what might be done if the VSWR is slightly outside the specifications.

7

TEES AND COUPLERS

Microwave systems are made up of many individual components joined by lengths of waveguide, coaxial line, or other transmission line. The components may be separate entities with flanges or connectors to fasten them to transmission lines and other components, or there may be two or more components built into one length of transmission line as a package. Although the individual components may have widely different functions, the problems in designing them are similar. A low standing-wave ratio is usually a requirement. This is achieved by the methods described in Chap. 6 regardless of the type of component being designed. Similarly, consideration of voltage breakdown and attenuation are the same for all components.

Microwave components and networks are usually designated by the number of *ports* entering and leaving the network. Thus, a microwave gadget which has one input arm and one output arm is a two-port network. In low-frequency circuits, a similar element would be called a four-terminal network, and it is thus evident that the number of ports is half the number of terminals. Either designation is technically correct, but it is easier to think of a waveguide opening as one port rather than two terminals. Tees and couplers are *junctions* or networks having three or more ports.

7.1. WAVEGUIDE H-PLANE TEE

When a short piece of waveguide is fastened perpendicular to the narrow wall of a straight length of waveguide, as illustrated in Fig. 7.1a, the three-port junction which results is called an *H-plane* tee, because of its similarity to an H-plane bend. The perpendicular arm is usually the input; and the other two arms are in shunt in the output. For this reason the tee is frequently called a *shunt* tee. (A simplified equivalent circuit

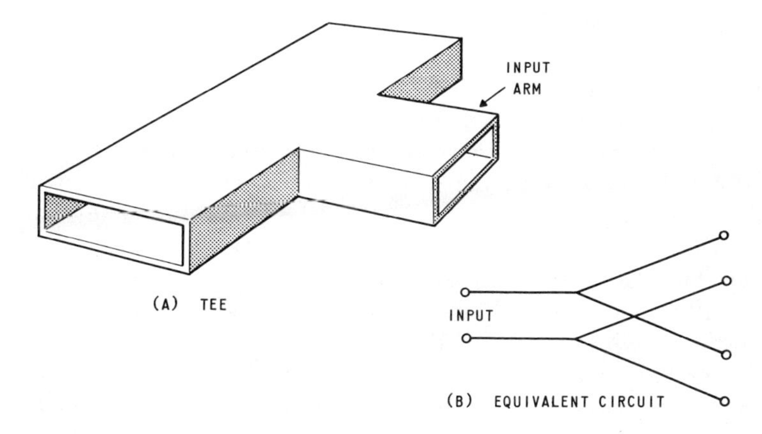

Fig. 7.1. H-plane tee.

is shown in Fig. 7.1b.) However, because of fringing fields inside the junction, a more exact equivalent circuit would include reactances in shunt and in series with the two output arms.

A simple method of matching a shunt tee is illustrated in Fig. 7.2. A slot is cut in the wall opposite the center of the input arm, and an

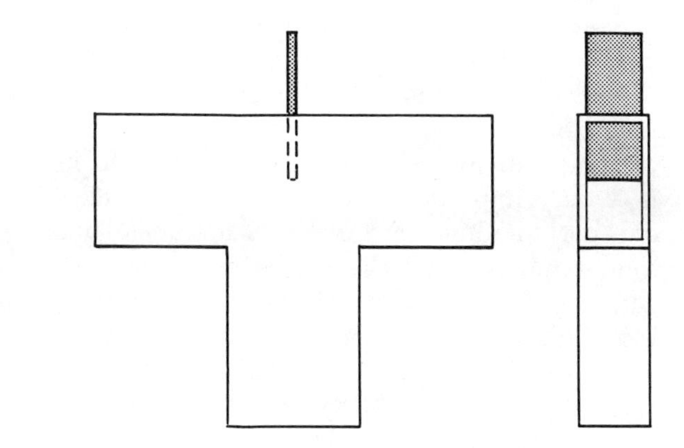

Fig. 7.2. Iris matching of tee.

adjustable iris which is the full height of the waveguide is pushed in. The thickness of the iris can be about $\frac{1}{10}$ to $\frac{1}{20}$ the height of the waveguide. The two output arms of the tee are terminated in matched loads, and the input is connected to a reflectometer and swept oscillator. As the iris is pushed into the function, a point will be found where the input arm is matched. The iris is soldered in place at this point. This method is satisfactory for small frequency ranges up to about five per cent.

If it were possible to cancel all the fringe reactances in the junction, the input arm would see two arms in parallel or half the impedance. The VSWR would be two to one but would be independent of frequency. It could then be matched with a suitable taper or transformers to cover a wide band of frequencies. The iris method (explained above) can be used, but instead of trying to achieve a good match when the iris is moved, the goal should be a two-to-one VSWR over the whole frequency band.

In place of an iris, a metal post can be used on the center line of the junction. The diameter and position of the post are varied until a two-to-one VSWR is obtained. Using a post in this fashion with the addition of a taper in the input arm, a VSWR under 1.10 has been obtained over a frequency band well in excess of 25 per cent.

It should be noted that it is impossible to match a three-port junction in all three arms unless the component is nonreciprocal. Nonreciprocity can be achieved only by the addition of some magnetic element and a magnetic field. In passive three-port components such as a tee junction, only the input arm is matched.

When a shunt tee is properly matched and terminated, the input arm presents a low VSWR. By symmetry, power fed into the input arm splits equally and in phase in the other two arms of the tee. By reciprocity, if two signals of equal amplitude and the same phase are fed into the two "output" arms of the shunt tee, they will combine and add in the "input" arm without mismatch. There is some coupling between the two output arms, and the amount of signal coupled is equal to and out of phase with the reflection at that arm so there is zero net reflection.

The principle of reciprocity, referred to in the preceding paragraph, is a very important one. In any passive electrical circuit, if a signal in one port produces a current in a load at another port, and the signal source and the load change places, the same current will be produced in the load. Applied to waveguide junctions this rule indicates that if a signal is applied to the first port of a multiport junction and produces an output signal which is p per cent of the input in a load at a second port, then, if the signal is applied at the second port, p per cent of it will appear at the same load placed at the first port.

It is possible to violate reciprocity by using ferrites and semiconductors in special circuits. However, it has been proved that in order to violate reciprocity, it is necessary to have a constant applied magnetic field somewhere in the circuit.

7.2. WAVEGUIDE E-PLANE TEE

If the input arm of the tee comes off the broad wall, as illustrated in Fig. 7.3a, the junction is called an *E-plane* tee, because of its similarity to an E-plane bend. The two output arms are in series as indicated in the simplified equivalent circuit of Fig. 7.3b, and thus the junction is commonly called a *series* tee.

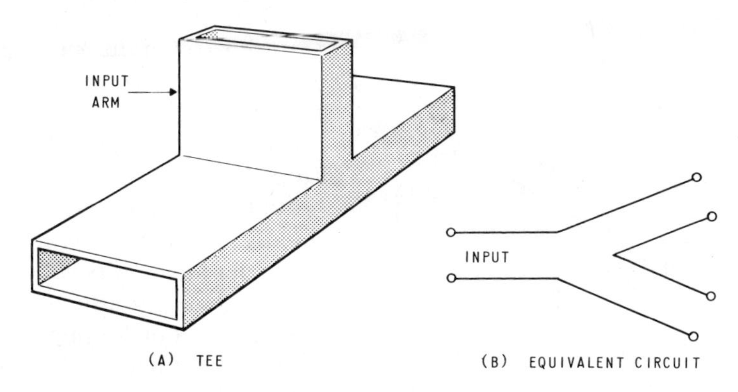

Fig. 7.3. E-plane tee.

An E-plane tee is most easily matched by using a susceptance in the input arm. This requires plotting on a Smith Chart and calculating the proper place to put an iris or post. The size of the iris can be determined from Fig. 6.10.

As in a shunt tee, the input arm of a properly matched and terminated series tee presents a low VSWR. Power fed into the input arm splits equally in the other two arms, but the outputs are 180° out of phase.

7.3. BIFURCATED WAVEGUIDE

Dividing the power with a good match over the full waveguide frequency range, is more easily accomplished by using a bifurcated waveguide as shown in Fig. 7.4. The cross section of the guide is divided into half-height waveguides by a septum extending all the way across the opening. The wide dimension of the two smaller waveguides is the same as the wide dimension of the original guide, but the heights at the start of the septum are just half the original height. To insure a good match, the septum ideally should come to a knife edge. Since the impedance of a waveguide with fixed wide dimension is directly proportional to the height, each half-height guide has half the impedance of the original waveguide. The equivalent circuit of the bifurcated junction is the same as that of a series tee; the two halves of the bifurcated section are in series as viewed from the input. There are no fringing fields. Thus, this junction is well-matched at all frequencies and needs no additional corrections. The two smaller guides in the bifurcated section may be tapered back to full-height waveguide, bent around corners to form a tee (illustrated in Fig. 7.4), or arranged in any geometry to meet system requirements.

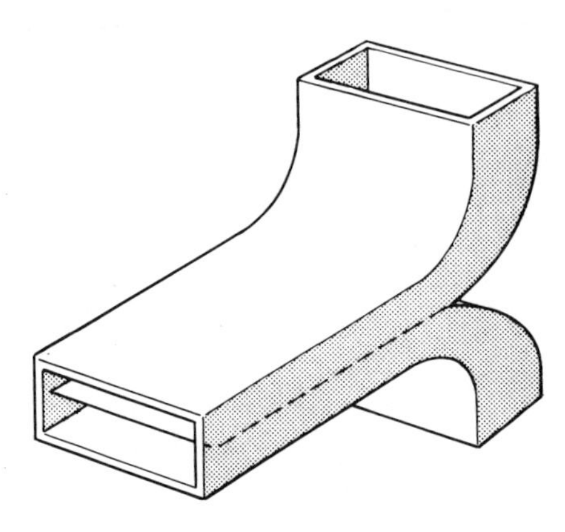

Fig. 7.4. Bifurcated waveguide.

It should be obvious that if the septum is not in the center of the waveguide, it will divide the power unequally. In fact, the power in each of the small waveguides will be proportional to the height at the opening. For example in an X-band waveguide which has a height of 0.400 inch a knife-edge septum is inserted which divides the height into 0.100 and 0.300 inch sections. The two small waveguides are then both tapered back to 0.400 inch height guide. (Incidentally the "knife edge" must also have a gradual taper since the septum cannot have zero thickness.) The power in the waveguide which started with 0.100 inch height will be

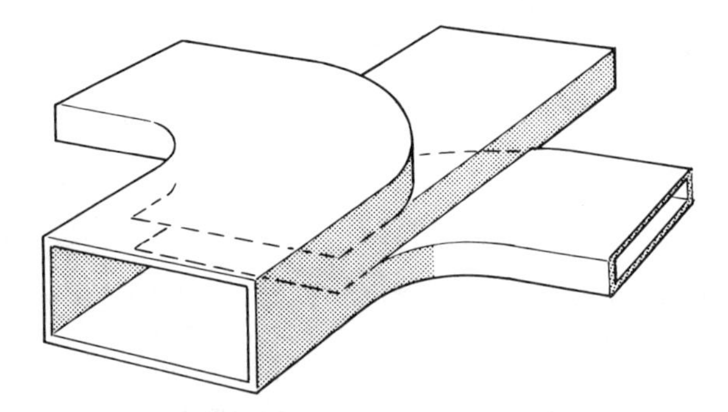

Fig. 7.5. Trifurcated waveguide.

one-fourth (0.100/0.400) of the input power. Similarly, the other part will have three-fourths (0.300/0.400) of the input power. It is possible to achieve any desired power division with a good match over a broad frequency range.

If more than one septum is used, dividing the original waveguide into three or more small height guides, the input power will split in proportion to the heights at the start of the septa. Figure 7.5 shows a trifurcated waveguide. Electrically, all the outputs are in series as seen from the input.

All the output signals are in phase in this type of junction and remain in phase as long as the new guides are not bent in opposite directions in the E-plane. However, after the bends illustrated in Fig. 7.4, the two signals are effectively 180° out of phase since one waveguide has been turned over with respect to the other. The H-plane bends illustrated in Fig. 7.5 do not invert the phase in this manner.

7.4. MAGIC TEE

A four-port junction which is a combination of an E-plane tee and an H-plane is called a *hybrid tee* or simply a *hybrid;* because of its unusual properties, which will be described below, it is sometimes called a *magic tee.* The tee is illustrated in Fig. 7.6. It has four arms or ports which have

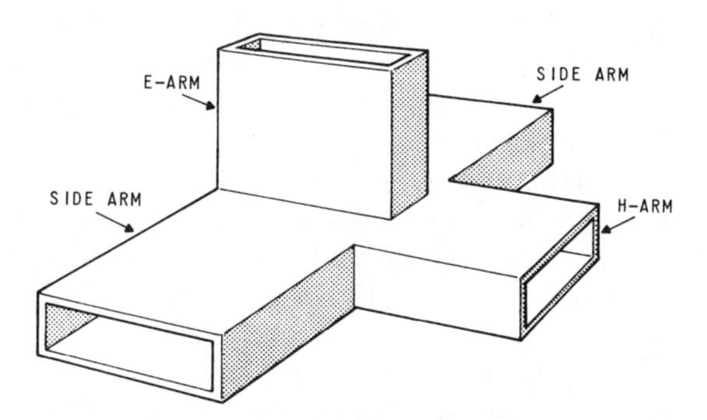

Fig. 7.6. Magic tee.

the names indicated in the figure. The colinear arms are called the *side arms,* implying that it is usually one of the other two arms that faces the viewer. The arm which makes an H-plane tee with the side arms is called the *H-arm* or *shunt arm.* The fourth arm makes an E-plane tee with the side arms and is thus the *E-arm* or *series arm.*

The shunt and series arms are cross-polarized; that is, the voltage vectors in these two arms are perpendicular to each other. Therefore, as long as there is nothing within the junction to rotate the polarization, there can be no coupling between these two arms. The E-arm "sees" only the side arms and, in fact, these three ports behave like an E-plane tee. Similarly the H-arm and side arms together behave like an H-plane tee.

To match the H-arm and the E-arm it is only necessary to follow the same procedures described in matching the H-plane and E-plane tees. Thus, the H-arm is matched by a septum which is opposite the center line of this arm. If the septum projects into the E-arm, it has little effect there since at that point it is perpendicular to the voltage. The E-arm can be matched with an iris or a post after the admittance has been plotted on a Smith Chart. Usually the H-arm is matched first so that the small effect of the septum projecting into the E-arm will be included in the Smith Chart plot.

Unlike the individual E-plane and H-plane tees, the side arms in the magic tee will be matched if the series and shunt arms are matched. Following the indicated matching procedure, first the H-arm is matched with matched loads on the side arms. Then the E-arm is matched with matched loads on the side arms. The side arms are now matched automatically if the shunt and series arms are terminated in matched loads.

The "magic" associated with this hybrid junction is the way that power divides in the various arms. If a signal is fed into the shunt or H-arm, power divides equally and *in phase* in the two side arms, with no coupling to the E-arm. When a signal enters the E-arm, it also divides equally in the two side arms, but this time the two halves are 180° out of phase, and there is no coupling to the H-arm. If power is fed into a side arm, it divides equally into the shunt and series arms and there is no coupling to the colinear side arm.

If a magic tee is perfectly made, the E- and H-arms are exactly perpendicular to the side arms. Now, if the E- and H-arms are perfectly matched, the power division and isolation indicated in the preceding paragraph are exact. However, in practice it would be fruitless to demand too high a degree of precision. For example, if the tee is perfect, but the matched loads (or matched antennas or other components) are not, a signal entering the H-arm would first split equally in the side arms, but the reflections from imperfect loads would enter both the H- and E-arms just as if they were new signals entering the side arms. If these two reflections from the side arms were exactly identical in phase and amplitude, they would cancel in the E-arm so that the isolation between the shunt and series ports would still be perfect. However, suppose that one side arm did have a perfect load, but the other was connected to an antenna with a VSWR of 1.2:1. The matched load would absorb all the power coming to it, but the voltage reflection from the antenna would be [from Eq. (2.25)]:

$$\Gamma = \frac{1.2 - 1}{1.2 + 1} = \frac{0.2}{2.2} = \frac{1}{11}$$

The power coming back from the antenna is the square of the voltage;

that is, $\Gamma_2 = 1/121$. Half of this goes back to the H-arm and half goes into the E-arm. This is approximately 24 decibels down from the original signal. It would obviously be unnecessary to specify isolation requirements of 50 decibels. For most applications 30 decibels is a reasonable specification for isolation and does not require difficult machining tolerances.

If signals are fed into both side arms, they will combine in the other two ports. The signals will combine in phase in the H-arm and 180° out of phase in the E-arm. For this reason the H-arm is sometimes called the *sum arm* and the E-arm is the *difference arm*. The original signals do not have to be in phase nor equal in amplitude. The output at the H-arm will be the vectorial sum of the two inputs at the side arms, and the output at the E-arm will be the vectorial difference between these two signals.

If the shunt and series arms of a magic tee are both terminated in movable short-circuiting plungers, the component is known as an *EH-tuner*. A signal entering a side arm divides equally in the two shorted arms, but the two halves of the signal are completely reflected. The relative phases of the reflected signals are determined by the positions of the plungers. The signals can thus be made to combine so that they produce any desired reflection coefficient in the input side arm. By the same token, the EH-tuner in front of a mismatched load can be made to produce a reflection which will exactly cancel the one from the load. This method of matching is very sensitive to frequency, but is useful for exact matching in the laboratory.

7.5. COAXIAL TEE

A three-port junction can also be built in coaxial line; it is usually called a tee even if the three arms are not perpendicular. As with any other three-port junction, it is impossible to match all three ports. The input port can be matched; the other two arms are outputs and are in shunt; the equivalent circuit is the same as the simple equivalent circuit for a waveguide shunt tee (shown in Fig. 7.1b).

Since the two output arms are in parallel, the impedance seen looking into the input arm would be half of the characteristic impedance of the lines if there is no matching. Obviously, the impedance can be matched by a quarter-wave transformer which has a characteristic impedance of $1/\sqrt{2}$, using the relationship of Eq. (6.2). This is illustrated in Fig. 7.7a. Alternatively, transformers could be put in each of the output arms, raising their impedances to two, so that when they are paralleled they will present a matched load to the unchanged input arm. The characteristic impedance of each of these transformers would be 1.414 ($= \sqrt{2}$), normalized. This case is illustrated in Fig. 7.7b.

Both of these methods of matching a tee are, of course, frequency sensitive since they use lengths of line which are quarter-wavelengths at only one frequency. By using transformers in all three lines it is possible to increase the bandwidth. (Figure 7.7c illustrates the choice of transformers which will give greatest bandwidth for a given VSWR.) The transformers in the output arm have a characteristic impedance of $\sqrt[4]{2}$,

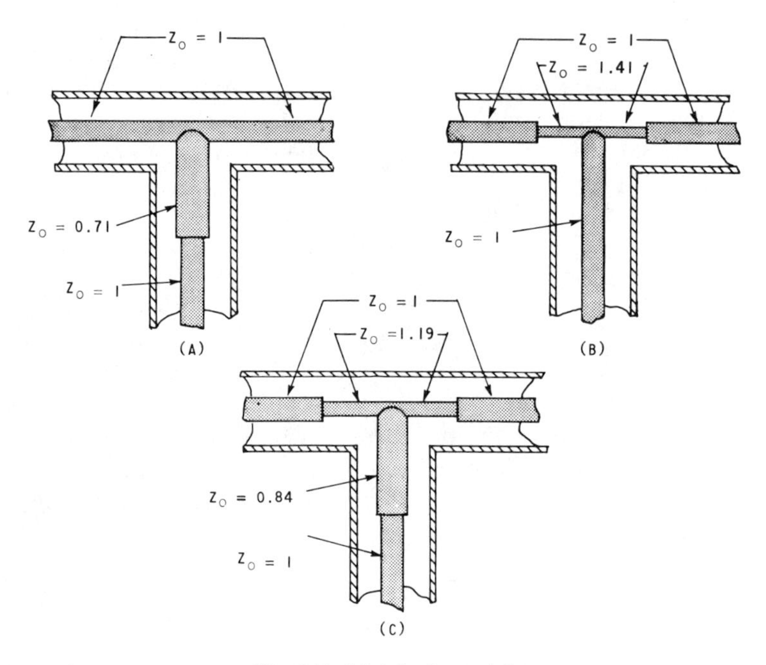

Fig. 7.7. Matched coaxial tees.

which is 1.19, and the quarter-wave transformer in the input arm has a characteristic impedance of the reciprocal of 1.19, which is 0.84. These are normalized values. The length of all the transformers is a quarter-wavelength at the center frequency. Theoretically, this arrangement will result in a VSWR under 1.2:1 over a two-to-one (octave) bandwidth.

If the transformers in the output arms are not identical, the power will divide unequally. If Z_1 is the characteristic impedance of the transformer in port one and Z_2 is that of the transformer in port two, using port three as the input port, the power will divide in this ratio,

$$\frac{\text{Power to arm one}}{\text{Power to arm two}} = \frac{Z_2^2}{Z_1^2} \tag{7.1}$$

The impedance seen at the junction looking toward arm one is Z_1^2 and looking toward arm two is Z_2^2. These two are parallel, and the parallel impedance is $Z_1^2 Z_2^2 / Z_1^2 + Z_2^2$. The value of a quarter-wave transformer in the input arm necessary to bring this impedance to unity is easily calculated.

7.6. DIRECTIONAL COUPLER

A directional coupler is a four-port component in which two transmission lines are coupled in such a way that the output at a port of one transmission line depends on the direction of propagation in the other. Figure 7.8 is a block diagram illustrating two transmission lines coupled in the junction—the line from port one to port two is coupled to the line

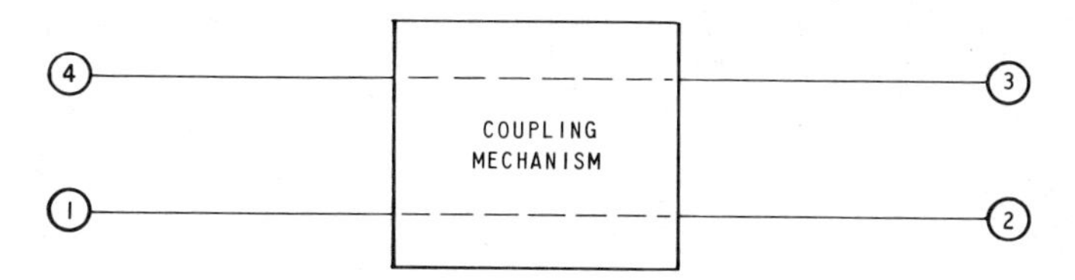

Fig. 7.8. Block diagram of directional coupler.

from port three to port four. In an ideal coupler, a signal entering port one will travel to port two, and a predetermined portion of this signal will appear at one of the other two ports. There will be zero output at the fourth port. If the main signal travels in the reverse direction, from port two to port one, the small coupled signal will appear at the port which was isolated in the first case. When a signal travels from port one to port two, the coupled signal can appear at either port three or port four, depending on the coupling mechanism used. If the signal in the side transmission line travels in the same direction as the main signal, the coupler is called a *forward* directional coupler. Thus, in Fig. 7.8, if the coupled signal output is at port three when the input is at port one, the coupler is forward. If the coupled output is at port four, it is a *backward* directional coupler.

The *coupling* of a directional coupler is the ratio of the input power to the coupled output power, expressed in decibels. Thus, if the power out of arm three is one-hundredth of the power into arm one, the component is a 20-decibel coupler. It should be noted that the power out of arm two must be reduced by the amount coupled out at arm three. Thus, in the 20-decibel example, since one per cent of the power has gone to arm three, the power out of arm two must be 99 per cent of the input. The reduction in power at arm two becomes greater as the coupling gets tighter. Thus, in a 10-decibel coupler, the power out arm two is only 90 per cent of the input, and in a three-decibel coupler it is 50 per cent.

All of the above examples assume that the coupler is perfect so that no signal comes out of the fourth arm. In practice, it is impossible to build a perfect coupler to cover even a narrow frequency band. A measure of the performance of a directional coupler is its *directivity*. This is defined as the ratio, expressed in decibels, of the coupling out of the coupling arm to the unwanted signal in the fourth arm. Thus, if the signal in the

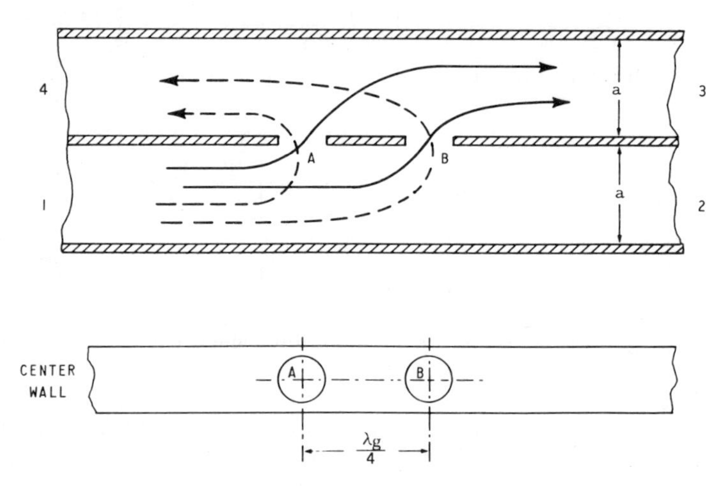

Fig. 7.9. Side-wall coupler.

fourth arm is one per cent of the coupled signal, the directivity is 20 decibels. The directivity can be more or less than the coupling; a 10-decibel coupler with 30-decibel directivity would be one in which the power out of the coupled arm was 10 decibels down from the incident, and the power out of the isolated arm was 40 decibels down from the incident or 30 decibels below the coupled signals.

Frequently only three of the four ports are used in a microwave circuit. In this case the unwanted port is usually terminated by a matched load built into it. The component then looks like a three-port network, but it is still a four-port network even though the fourth port is concealed. The built-in load must have an excellent VSWR since any reflections from it will appear almost entirely at the coupling arm.

In order to achieve this directional characteristic, it is necessary to provide two or more coupling mechanisms between the two transmission lines. The arrangement of couplings is such that in one direction of propagation in the secondary line all the coupled signals add in phase, while in the other direction their phases and amplitudes cancel exactly. A simple example is the waveguide two-hole side-wall coupler shown in Fig. 7.9. The two waveguides are placed side by side with their narrow walls in contact. In practice, the common wall is reduced from double thickness to single thickness by cutting one narrow wall off of one guide before joining the two. The solid lines indicate the signal paths from

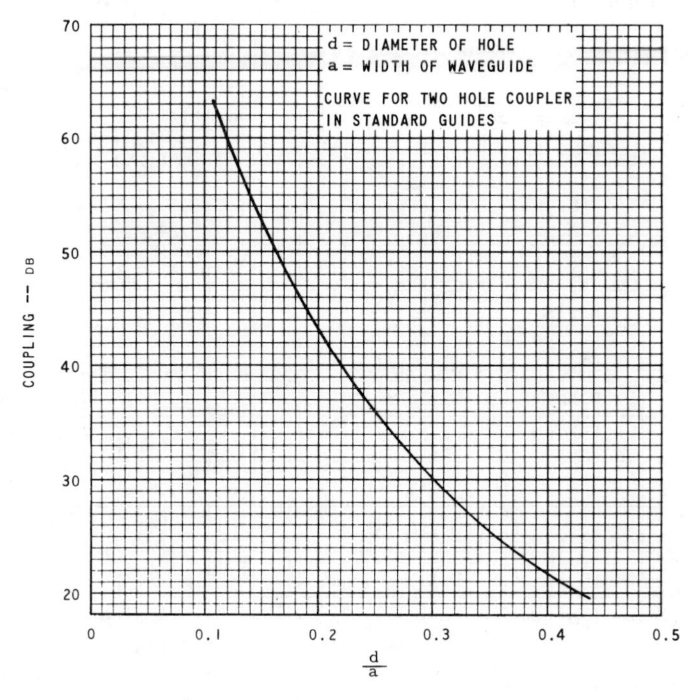

Fig. 7.10. Design chart for side-wall coupler.

port one to port three through the two coupling holes. Since both paths are the same length, there will be coupling at port three. The dotted lines show the signal paths from port one to port four. If the two holes are a quarter of a guide wavelength apart, these two paths will be 180° out of phase and there will be no coupling out of arm four. The amount of coupling out of arm three is a function of the hole size. A design chart of coupling as a function of hole size is given in Fig. 7.10.

The two holes in the coupler described above must be a quarter-wavelength apart to have perfect directivity. Obviously, since this happens at only one frequency, the directivity varies with frequency. The coupling also varies with frequency since it depends on the size of the hole as compared to a wavelength. The performance of any type of coupler can be improved by using more holes or coupling mechanisms. In general, the greater the number of coupling mechanisms, the wider the frequency bandwidth and the greater the directivity.

The two waveguides can have their broad walls in common instead of their side walls. The coupler uses holes spaced a quarter-wavelength apart in the direction of propagation and is a forward coupler. Multihole top-wall couplers of this type are used for reflectometers since they have

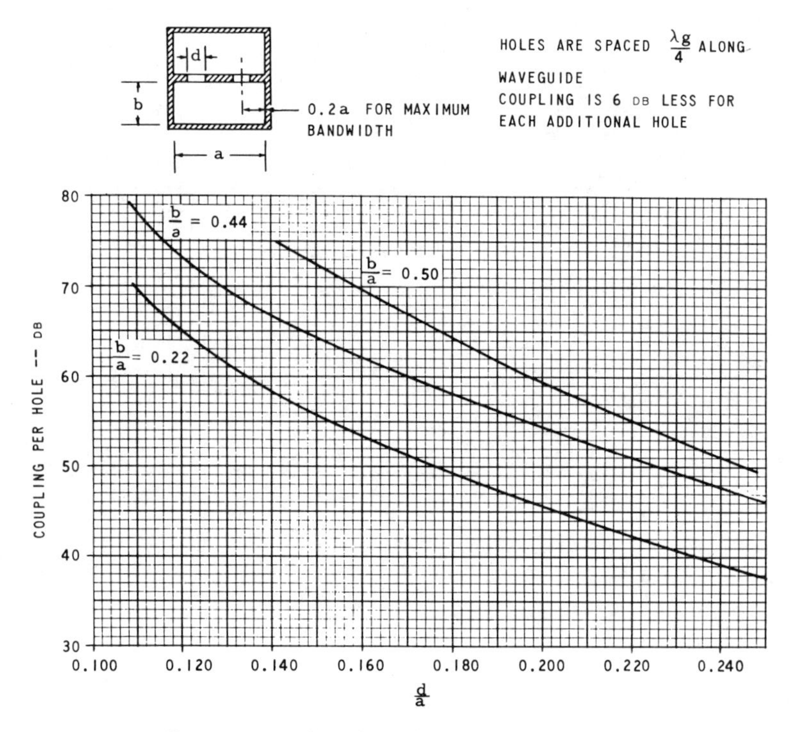

Fig. 7.11. Design chart for top-wall coupler.

excellent directivity and cover a broad frequency band. A design chart for this type of coupler is shown in Fig. 7.11.

Another type of waveguide directional coupler consists of two waveguides crossing at right angles with one broad wall in common. The coupling holes are usually slots or arrangements of crossed slots, rather than round holes. This type of coupler is not usually found in laboratories since it does not have as broad a band as the top-wall coupler, but it does find application in systems where the crossed waveguides present a favorable geometry to other parts of the system.

The coupling mechanisms do not necessarily have to be holes in a common wall. They can just as well be other transmission paths. Figure 7.12 shows a schematic of a branch-guide coupler in which the coupling mechanisms are quarter-wavelength sections of transmission line. The coupling and match of a branch-guide coupler are determined by the characteristic impedances of the branches. Although Fig. 7.12 shows three branches, the number is not restricted, and, as before, the more branches there are, the greater will be the bandwidth and the directivity.

The branch-guide configuration can be used with coaxial line or strip line as well as with waveguides. In coaxial lines the length of the branches

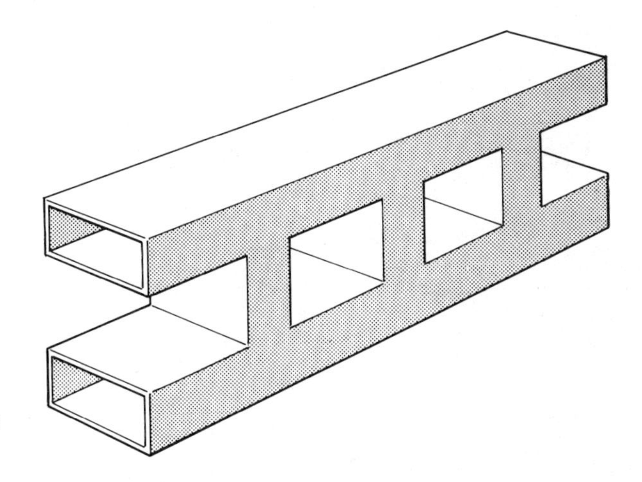

Fig. 7.12. Branch-guide coupler.

and the spaces between branches is a quarter-wavelength. In waveguide it is a quarter of a guide wavelength.

Another type of strip-line directional coupler depends on electromagnetic coupling. The "inner" conductors of two strip lines run parallel to each other for a distance of a quarter-wavelength between common ground planes. The amount of coupling depends on the spacing between the parallel conductors, and the match depends on their impedance. The directivity is excellent at the design frequency, where the coupling region is a quarter-wavelength, and is quite good over an octave bandwidth centered at this design frequency.

7.7. HYBRIDS

The magic tee, described in Sec. 7.4, is called a hybrid, possibly because it is a combination of an E-plane tee and an H-plane tee. The magic tee can also be considered a three-decibel coupler since the power entering any arm splits equally in two other arms, and the fourth arm is isolated. Each of the two output arms, of course, has half the input power, which is a drop of 3 decibels. The designation *hybrid*, or *hybrid junction*, or *hybrid coupler* is applied to any 3-decibel coupler, whether it is used in waveguide, coaxial line, or any other form.

Hybrids fall into two broad classes. In the class for which the magic tee is typical, the two outputs are either in phase or 180° out of phase, depending on which arm is used as the input. In the second class of hybrids, the two outputs have a 90° phase difference, regardless of which is the arm input.

If the directional couplers described in the preceding section can be made to couple half the power into the auxiliary line by providing a

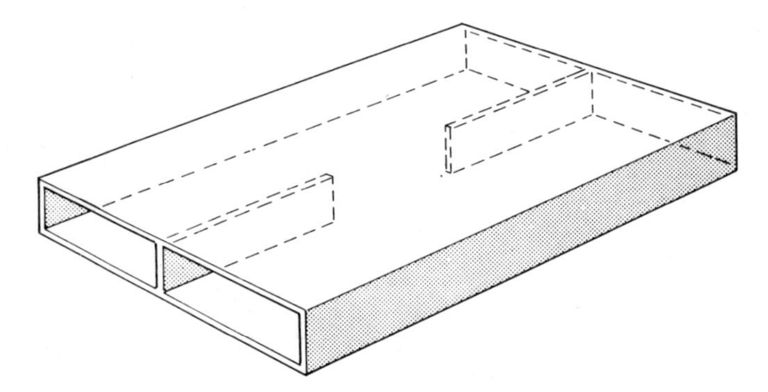

Fig. 7.13. Short-slot coupler.

sufficient number of coupling paths or enlarging the coupling paths or holes, they would be hybrids. This can be done with some of the configurations, particularly the side-wall coupler, top-wall coupler, and branch coupler. In all of these, the signals in the two output arms are 90° out of phase, or in *quadrature*.

The side-wall coupler can be made into a 3-decibel coupler by using a large number of holes. However, the same effect can be achieved simply by completely removing a section of the common wall between the two waveguides and adding some reactive elements for matching. The result, illustrated in Fig. 7.13, is a compact hybrid which is called a *short-slot* coupler. The operation of this component can be explained by considering the section containing the slot to be a double-width waveguide. When the input signal enters this section, the sudden discontinuity excites higher modes. In this region both the $TE_{1,0}$ mode and the $TE_{2,0}$ mode can propagate. Part of the signal travels in each of these modes in all directions. If the path lengths are correct, the two signals (both modes) will cancel in the isolated arm and combine in the two output arms to become a 3-decibel coupler. The two outputs are 90° out of phase.

Multihole top-wall hybrids are used in the laboratory because they have high directivity. As with the short-slot coupler, a similar compact top-wall hybrid can be formed by removing a large portion of the common wall as shown in Fig. 7.14. This hybrid also has outputs which differ by 90°.

Branch guide hybrids can exist in an infinite number of forms. Some of these are shown schematically in Fig. 7.15. The lines represent transmission lines and can be waveguides or coaxial lines. The length between

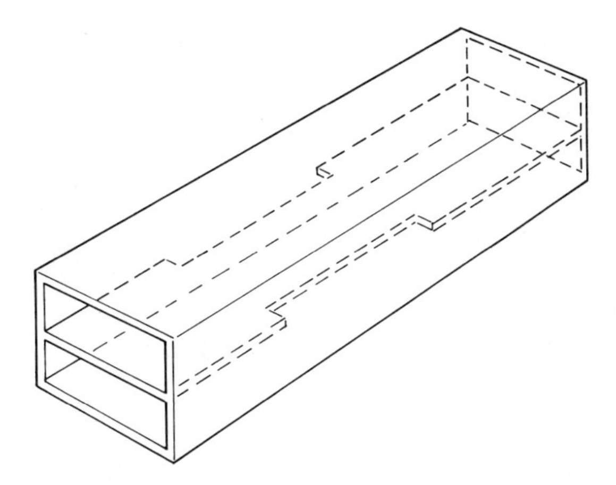

Fig. 7.14. Top-wall coupler.

intersections is always a quarter-wavelength in the transmission line. The numbers beside each section of line represent the normalized characteristic impedance of that particular section. In all the configurations shown, a signal entering either arm at the left will split equally in the two arms at the right with a 90° phase difference.

The bandwidth of a hybrid is the frequency range within which the unit meets specifications. At the design frequency, ideally there is an exactly equal power split—perfect isolation in the fourth arm and a perfect match in the input arm. Off frequency, there is a difference in power at the two output arms, the isolation decreases, and the VSWR in the input arm increases. The hybrids shown in Fig. 7.15 are arranged in order of bandwidth, with the narrowest at the top and the widest bandwidth hybrid at the bottom. The square hybrid at the top is a comparatively narrow-band component. At the band edges of a 12 per cent band, its VSWR is up to 1.26, the isolation in the fourth arm is down to 19 decibels, and the powers in the two output arms differ by a quarter of a decibel. For the same bandwidth, the four-branch hybrid at the bottom, in contrast, has a VSWR of 1.01, as much as 45-decibel isolation, and only a tenth of a decibel difference in output.

Another type of hybrid which can be built in either waveguide or coaxial line is called a *rat-race*. As shown schematically in Fig. 7.16, it consists of a circle that is one and a half wavelengths long and has a

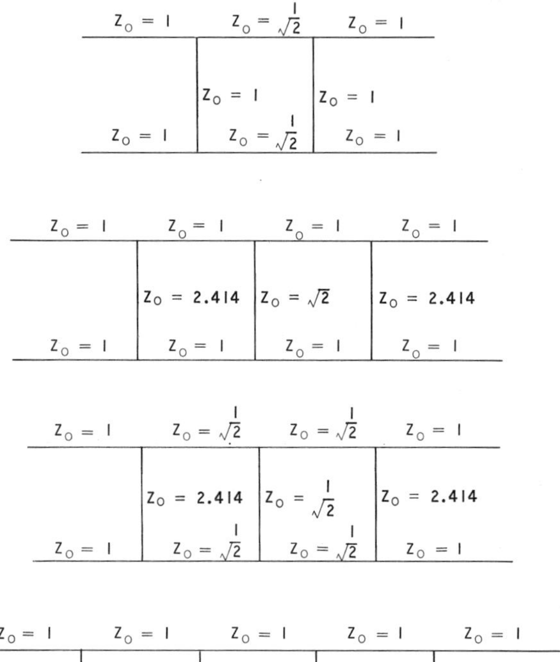

Fig. 7.15. Branch-guide couplers.

normalized characteristic impedance of 1.414. The four ports are connected to this circle with spacing as shown. The phase characteristics and operation of this hybrid are similar to those of the magic tee. With power in port one, half appears at port two and half at port four, with 180° phase difference between ports. Power in port three also splits between two and four, but in phase. There is isolation between port one and port three and also between port two and port four. By symmetry, it is evident that power in port two splits in phase in arms one and three, while power in port four splits in the same arms but out of phase. In bandwidth, the rat-race lies between the two three-arm branch hybrids of Fig. 7.15.

Hybrids can be used to combine two signals as well to divide one signal in two. By reciprocity, if two signals enter two output arms in the correct phase relationship, they will combine into what is usually an input arm. Even if the phases are not correct, they will combine in two other arms

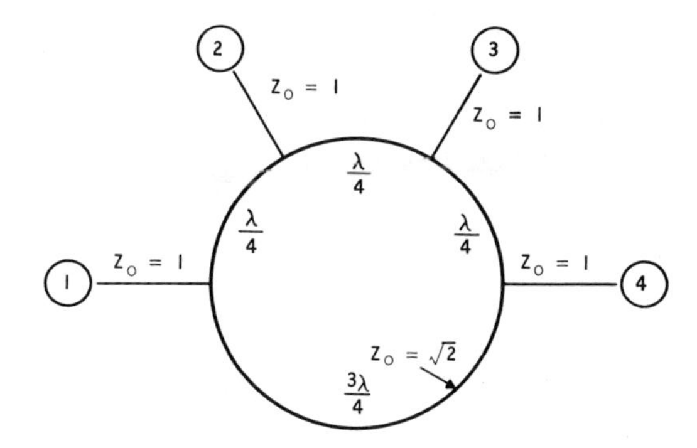

Fig. 7.16. Rat race.

in some predetermined fashion. For example, in the rat-race, as in the magic tee, if signals enter arms one and three, they will add algebraically in arm two, and their algebraic difference will appear in arm four.

QUESTIONS AND PROBLEMS

7.1. Explain a method of matching the input port of a shunt-T to the collinear arms.

7.2. Draw the low-frequency equivalent circuits for both H- and E-plane T-junctions. Point out the phase relationships in the arms.

7.3. What are the points of similarity between the circuit shown here and the "magic tee"?

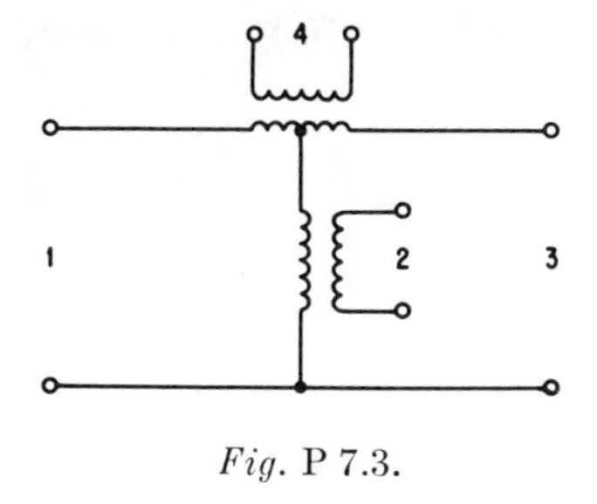

Fig. P 7.3.

7.4. Derive the formula stated in Eq. (7.1).

7.5. Show how the EH-tuner might be considered the low-frequency equivalent of the "L" matching network.

7.6. Define the terms "coupling" and "directivity" as they are used in connection with directional couplers.

7.7. You have two identical 20-db directional couplers in a guide to sample the incident and reflected powers. The outputs of the two couplers are 3.0 Mw and 0.1 Mw respectively.
   a. What is the VSWR in the main guide? *ans.* 1.45
   b. What is the power being dissipated in the load?
   c. What is the value of the reflected power? *ans.* 10 Mw

7.8. What is the coupling coefficient of the coupler shown in the diagram below? What is the return loss?

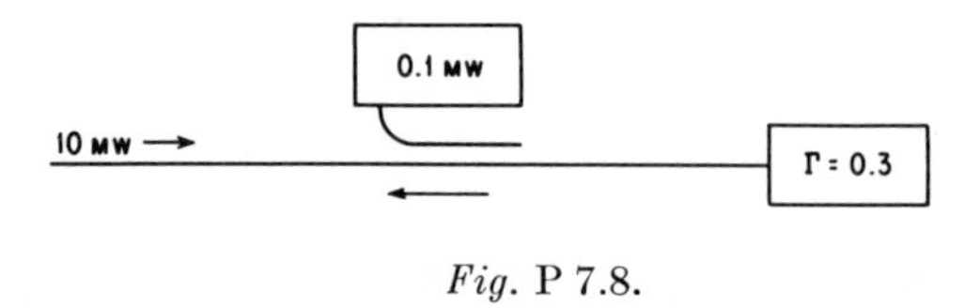

Fig. P 7.8.

7.9. By means of diagrams show how a "magic tee" could be used to check if two loads have identical values. What limits the use of this system as a means of checking the value of unknown impedances?

7.10. Give a quantitative example to show the possible error caused in checking the power into a load with a 20-db coupler that has 40-db directivity.

8

MICROWAVE COMPONENTS

Every microwave system consists of many components which are joined by sections of transmission line. These components direct the microwave energy, control its phase and amplitude, transform it to other forms of energy, or sample the energy so that it can be measured. This chapter will discuss some of the passive components which are frequently used but are not so complicated that they need complete chapters of their own.

8.1. MATCHED LOADS

Many components used in microwave circuits require a load which completely absorbs all the incident power. A typical example is a directional coupler in which only three of the four ports are used. The device which has an excellent match (zero reflection) and complete absorption of all the power incident on it is called a *matched load* or a *matched termination.*

Matched loads are necessary in the laboratory when other microwave components are being developed and tested. In order to check the match at a discontinuity, it is necessary that no other source of reflection be in the circuit. This can be accomplished by placing a matched load beyond the discontinuity to absorb all the power that passes it.

A simple form of matched load in waveguide is a piece of resistance card placed in the waveguide parallel to the electric field, as shown in Fig. 8.1. The front of the card must be tapered so that it presents no discontinuity to the microwave signal. If the card is long enough to absorb almost all of the power, the reflections from the far end will be sufficiently small so that the net reflection is negligible. Thus, if the card has 20 decibels of attenuation in one direction, there is a two-way attenuation of 40 decibels. Even complete reflection from the far end, down 40 decibels, would produce a VSWR at the front of 1.02, which is negligible for most applications.

Even with a taper, the front end of the card is never a perfect match, because the backing material on which the resistive film is deposited has

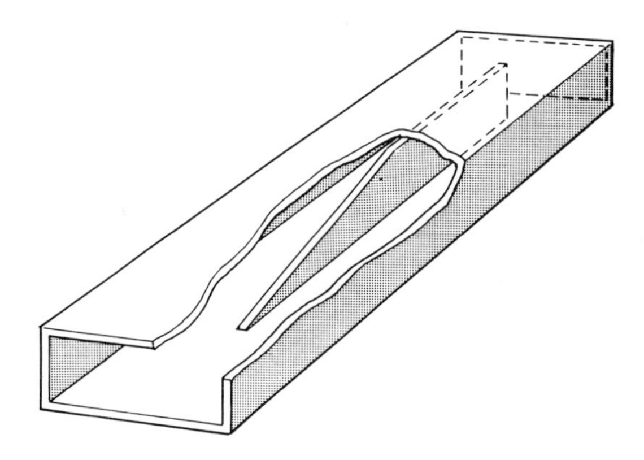

Fig. 8.1. Card load.

a finite thickness and produces a discontinuity. Whenever this small reflection is important, the load should be slid along the waveguide. If the load is perfect, there will be no variation in the VSWR as it is moved. If there is a change, then the maximum and minimum values of standing-wave ratio should be noted. The VSWR of the load and that of the rest of the circuit can then be determined from Eqs. (2.43) and (2.44). Usually the smaller value will be that of the load.

If the card is on the center-line of the waveguide, it will produce the greatest attenuation, since it is at the point of maximum field. Any discontinuity at the front will also be a maximum. Thus, if the card is long enough, it can be moved closer to a side wall of the waveguide, where the reflection from the front will be lessened.

Any lossy material can be used as a matched load in a waveguide as long as provision is made to avoid a reflection from the front end. Figure 8.2 shows some configurations of solid loads which fill the waveguide cross section. In each case a taper or tapers must be used to avoid the first reflection. The knife edge of the taper in Fig. 8.2a is in the center of the waveguide and will produce a greater discontinuity than the horizontal taper of Fig. 8.2b. For minimum reflection, however, the taper should begin near the side walls where the voltage is zero, as shown in Figs. 8.2c and 8.2d. These loads, like the cards, can be slid along the guide to different positions.

Materials commonly used for solid loads are lossy dielectrics, dielectrics loaded with carbon or powdered metal, wood, sand, or, in fact,

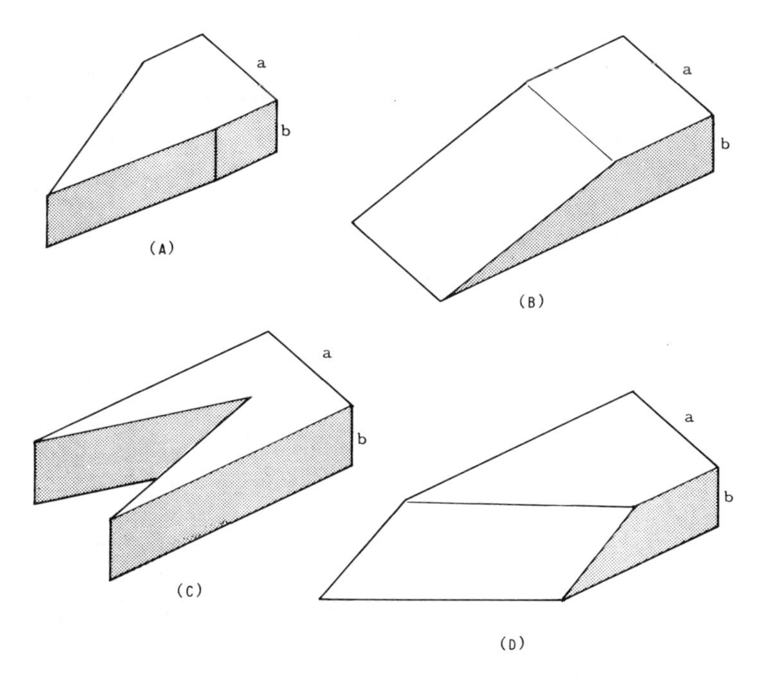

Fig. 8.2. Solid waveguide loads.

anything which is not a good conductor. Where high temperature or high power is a consideration, it is necessary to use a material which will not melt at the anticipated temperature. Powdered iron in a silicon base is frequently used. To remove the heat from the waveguide as rapidly as possible, radiating copper fins may be soldered to the outside. For additional cooling, forced air may be blown against the fins.

When a very compact load is needed, as in the closed arm of a directional coupler, a solid load is used, but, without the taper. Instead a step is used as shown in Fig. 8.3. The step is similar to a quarter-wave transformer and produces a narrow-band match. In effect, the two discontinuities produce reflections which are equal and 180° out of phase, and therefore cancel.

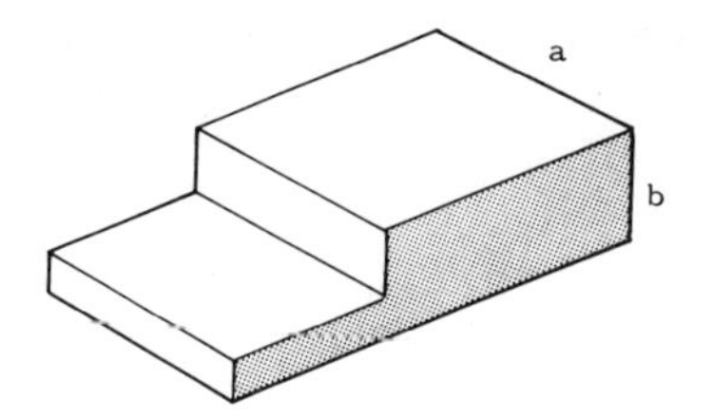

Fig. 8.3. Stepped load.

In round waveguide, using the $TE_{1,1}$ mode, the same considerations apply. A card may be used parallel to the E-field just as in rectangular waveguide. Solid plugs with tapers are also effective. Similarly in coaxial line, cards and solid loads are effective. Since the field is greatest next to the inner conductor, a solid load in coaxial line should have a taper which begins at the outer conductor. As with rectangular waveguides, fins are also used with round guides and coaxial lines to provide cooling in high power applications.

8.2. FIXED ATTENUATORS

Any of the loads described above is, of course, an attenuator which absorbs almost all of the power. If the load is shortened and a matched taper placed on the output as well as the input, a fixed amount of power can be absorbed by the load while the rest is transmitted. As before, attenuators can use cards or solid lossy materials. The amount of attenuation is the reduction in the signal expressed in decibels. Thus, if the power output of an attenuator is one-fourth of the input, the attenuator has a value of six decibels and is called a 6-db *pad*.

The attenuation of a fixed pad varies with frequency for two reasons: at higher frequencies, there are more wavelengths in the lossy material, and also the loss in the material itself is dependent on frequency. It is necessary, therefore, to specify the frequency as well as the attenuation of a fixed attenuator.

8.3. VARIABLE ATTENUATORS

It is frequently necessary to vary or otherwise control the amount of signal power in a given circuit. For example, a crystal detector has a reflection coefficient which varies with the incident power. To determine this relationship the power incident on the crystal is varied during the measurement procedure—usually by varying the amount of attenuation between the signal generator and the load. Another application of a variable attenuator is the control of local oscillator power in a microwave receiver.

Since a piece of resistance card in a waveguide intercepts most power when it is at the center of the guide and practically no power when it is next to the side wall, it is possible to build a very simple variable attenuator making use of this property. The resistance card is supported by two thin horizontal rods which protrude through small holes in a narrow wall. A knob and gears control the movement of the card from the wall to the center. The horizontal rods introduce negligible reflection since they are perpendicular to the electric field. The card, of course, must be tapered at both ends so that it produces no mismatch.

For more accurate attenuators, the resistance card is replaced by a piece of glass which has a metallized film on one side. The method of operation is the same, but the metallized glass is more stable with time and less subject to environmental change than the resistance card. A calibrated dial may be used to preset any required attenuation level.

Another form of variable attenuator uses a piece of slotted waveguide. Since the slot is on the center line of the broad wall, there is no radiation loss. The attenuation is varied by inserting a piece of resistive card or metallized glass into the slot. With the resistive material extending all the way across the waveguide, there is maximum attenuation. This decreases continuously while the card is withdrawn. As before, suitable mechanical linkages and a calibrated dial may be added so that the attenuation may be read directly.

The variable attenuators discussed thus far present two problems. First, they are frequency sensitive, just as fixed pads, so that it is necessary to calibrate them at each operating frequency if it is required to know the amount of attenuation. In the second place, they produce a phase variation with the attenuation variation. This is caused by the dielectric or glass member, depending upon its position, intercepting a varying amount of electric field.

A form of variable attenuator which is not frequency sensitive and which has a constant phase shift through it, regardless of the attenuation, is illustrated in Fig. 8.4. It consists of three sections of circular waveguide, each containing a sheet of resistive material. These resistance cards are long enough to absorb completely any signal whose electric field is parallel to them. Their center section is rotatable, but the two end sections are fixed with their resistance sheets perpendicular to the electric field of the input and output signals. In practice, gradual tapers from rectangular to round waveguide are used at each end.

It should be obvious, that if the center section has its resistance card parallel to the other two, there will be negligible attenuation, since they are all perpendicular to the electric field. On the other hand, if the center section is parallel to the electric field (perpendicular to the two fixed cards), there is complete absorption. At any angle, the amplitude of the signal that emerges is

$$A = E \cos^2 \theta \tag{8.1}$$

where E is the amplitude of the input signal, and θ is the angle between the fixed card and the rotating vane. The attenuation is thus,

$$\text{db} = 20 \log \frac{E}{E \cos^2 \theta} = 20 \log \sec^2 \theta \tag{8.2}$$

This is independent of frequency so that it is possible to calibrate the dial directly in decibels.

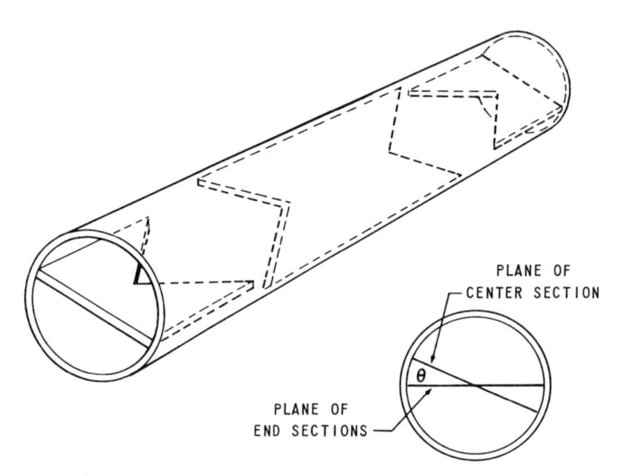

Fig. 8.4. Rotary-vane attenuator.

The quantities specified in a variable attenuator are the maximum attenuation, minimum insertion loss, and maximum VSWR. If the attenuator is designed for a band of frequencies, these quantities are usually determined for the two frequency limits and the center frequency.

8.4. CUT-OFF ATTENUATOR

When a waveguide is below cut-off, the microwave signal entering it is reflected as if from an open circuit. However, it is not complete reflection, and if the below cut-off section is not too long, some signal, although greatly attenuated, will appear at the far end. The signal entering the cut-off section of guide does not change phase but decreases exponentially with length. The amount of attenuation is given by

$$\alpha = 2\pi \sqrt{1/\lambda_c^2 - 1/\lambda^2} \tag{8.3}$$

where α is the attenuation in nepers per unit length, and λ_c and λ are the cut-off and free-space wavelengths, respectively. The attenuation is consequently directly proportional to the length of the cut-off section and can be determined from the physical dimensions of the waveguide.

When the signal wavelength, λ, is much less than the cut-off wavelength, the second term under the square-root sign in Eq. (8.3) becomes negligible. The relationship then becomes

$$\alpha = \frac{2\pi}{\lambda_c} \tag{8.4}$$

Eq. (8.4) indicates that as long as the signal wavelength is much longer than the cut-off wavelength, the attenuation is independent of frequency and dependent only on the dimensions of the cut-off section.

Although coaxial line does not have a low-frequency cut-off, a cut-off attenuator can be built by putting a section of circular waveguide in a

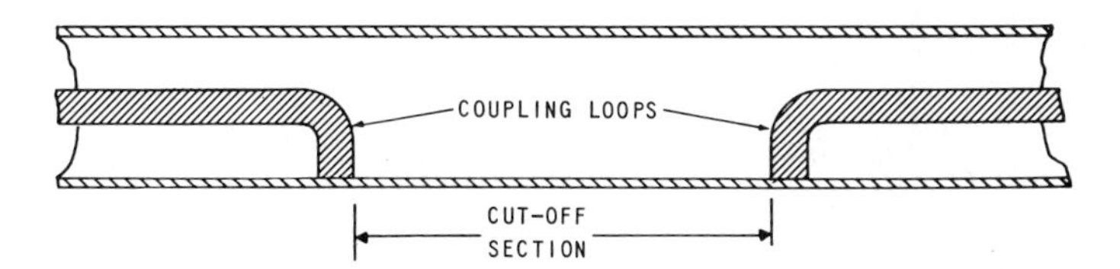

Fig. 8.5. Coaxial cut-off attenuator.

coaxial line. This is illustrated in Fig. 8.5. The input and output coaxial lines are coupled to the circular waveguide by loops which excite the $TE_{1,1}$ mode. The round guide diameter is chosen so that λ is much less than λ_c. The attenuation is then given by Eq. (8.4) and is proportional to the spacing between the loops. A convenient form for this attenuator has a mechanical linkage so that the spacing between the loops may be varied, and the attenuation may be read directly from a calibrated dial.

8.5. VARIABLE PHASE SHIFTERS

Just as a resistance card has a greater effect in the center of the guide than near a side wall, so also will a piece of low-loss dielectric intercept a greater portion of the microwave signal when it too is placed at the center. With the waveguide completely full of a dielectric, the guide wavelength is shortened as indicated by Eq. (4.12). This means that in a given length of dielectric-loaded guide there are more wavelengths than in the same empty guide. Thus, introduction of the dielectric produces a change in electrical length, which is another way of saying a change in phase. When the waveguide is partially full of dielectric, there is again a change in phase, although not as great a change as occurs with complete filling. The phase change also depends on the amount of interaction between the dielectric slab and the signal. Thus, it may be assumed that the phase will vary as a slab is moved from the point of maximum interaction (at the center of the waveguide) to the side wall. This is the basis for a variable phase shifter which is similar to the first variable attenuator described above. Another form of variable phase shifter, also like an attenuator, consists of a slab which is moved into the waveguide through a slot in its broad wall.

In designing variable phase shifters, it is important to taper the ends of the dielectric slab so that there is no loss due to reflections. The material is assumed to be lossless but there is always some loss present. Thus, it is necessary to measure the insertion loss as a function of phase shift. The quantities usually specified are the maximum phase change, the maximum VSWR, and the maximum insertion loss.

8.6. TRANSITIONS

In microwave circuits it is frequently necessary to change from one type or size of transmission line to another. To do this without losing power in reflections from the discontinuity, a well-matched transition unit must be used. Any component which accomplishes the change, while keeping VSWR within specifications over the required frequency band is called a *transition* or a *transducer*.

Transitions can be simple tapers from one size coaxial line to another. If the taper is long enough on either the inner or outer conductor or both, there will be very little reflection. Thus a taper can be used to change from one impedance coaxial line to another, or from one size to another, without change of impedance. If it is necessary to change from an air-filled coaxial line to one which is dielectric-loaded without a change in dimensions, a taper can be used on the dielectric.

Transitions from one rectangular waveguide to another can be accomplished by smooth tapers, as long as the frequencies of interest are above cut-off for both guides. The same sort of gradual taper can also be used to change from rectangular to square, rectangular to circular, or circular to square. Some of these transitions are difficult to fabricate by ordinary methods so an electro-forming process is used. First, a mandrel shaped to the inside of the required transition is made of aluminum or plastic. Then copper is deposited electro-chemically to a suitable wall thickness, and the mandrel is removed. Flanges can be soldered on the ends of the electro-formed transition to join with the proper waveguides.

When the two waveguides have the same impedance as determined from Eq. (4.10), they will have only a small reflection when they are butted up against one another even if they are different shapes. Thus, a 500-ohm circular waveguide in contact with a 500-ohm square waveguide will have a VSWR under 1.10. The small reflection is caused by the capacitive discontinuities at the juncture, but there is no impedance change giving rise to a large mismatch. If the waveguides are radically different in size, it may be necessary to add reactive elements to achieve a low VSWR transition.

Coaxial lines also can be butted together with small mismatch when they have the same characteristic impedance. A frequent form of coaxial transition is from a 50-ohm air-filled line to a 50-ohm dielectric-loaded line. A butt transition in this case is usually quite satisfactory.

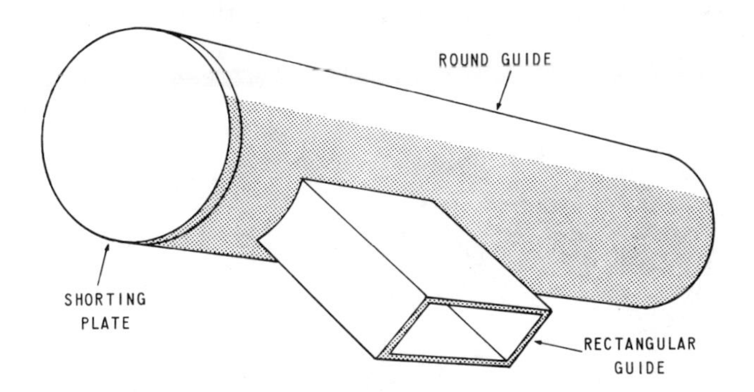

Fig. 8.6. Rectangular-to-round transition.

When the characteristic impedances are different in the two coaxial lines or two waveguides to be joined, a satisfactory transition can be a quarter-wavelength transformer, as was discussed in Sec. 6.1. In coaxial line, the transformer can be air-filled or dielectric-loaded regardless of the state of the individual coaxial lines to be joined. In waveguide, Eq. (4.10) should be used to determine suitable dimensions for the transformer. An interesting case occurs when the two waveguides to be joined have the same wide dimension, a, but different heights, b_1 and b_2. Since λg is the same for both, a quarter-wavelength transformer can be made with the same wide dimension and with height $b = \sqrt{b_1 b_2}$. This can be verified from Eq. (4.10).

A transition from rectangular to circular waveguide can also be made with the waveguides perpendicular to each other, as in Fig. 8.6. The rectangular waveguide feeds into a side of the round waveguide. A short-circuiting plunger is adjusted in one side of the round guide so that reflections from it reinforce the signal going in the other direction in the round guide. Final matching is done by an inductive post or iris.

If the short circuit of Fig. 8.6 is replaced with a rectangular waveguide which is cross-polarized with respect to the first rectangular guide, there will be no coupling between the two rectangular guides. The guide on the end makes a butt joint with the round waveguide which can be matched independently of the other transitions. A $TE_{1,0}$ mode signal from either rectangular guide will excite a $TE_{1,1}$ signal in the round guide which will propagate without respect to the other rectangular guide. However, the two $TE_{1,1}$ signals are cross-polarized. Reciprocally, a signal in the round guide travelling toward the transitions will enter the rectangular guide of the proper polarization. If the circular waveguide signal is at an acute angle with the two directions of polarization, it will split, and the power entering each arm will be proportional to the square of the cosine of the angle between the polarization in the circular guide and that in the rectangular. This transition is called a *dual-mode transducer* and is shown in Fig. 8.7.

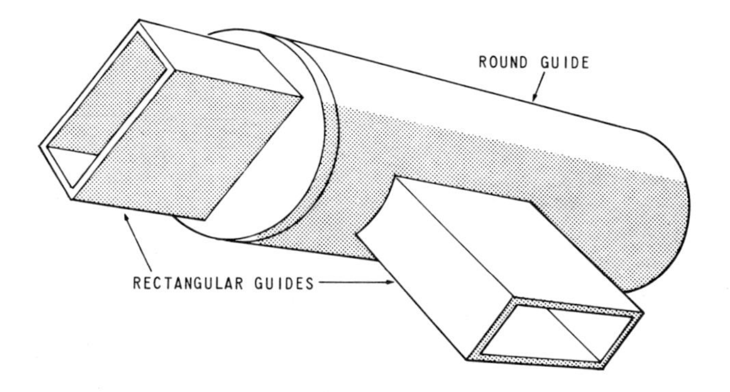

Fig. 8.7. Dual-mode transducer.

All the waveguide-to-waveguide transitions discussed thus far are junctures of two waveguides operating in the dominant mode. It is also possible to match a juncture of one waveguide in its dominant mode to another in a higher mode. A common transducer of this type joins a rectangular waveguide operating in the $TE_{1,0}$ mode to a circular guide in the $TM_{0,1}$ mode. But some of these mode transducers have no external change in dimensions. Thus, it is possible to change from the $TE_{1,0}$ to the $TE_{2,0}$ mode in rectangular guide, assuming the guide is large enough to support both modes. In fact, it is possible to change from any mode to any other without altering dimensions, as long as both modes can be supported. However, extreme care must be taken to prevent the dominant mode from being excited where it is not wanted. If it does appear, it will draw energy from the desired mode.

8.7. WAVEGUIDE-TO-COAX TRANSITIONS

Almost every microwave circuit or system has both waveguides and coaxial lines (or components) built into both types of transmission line.

At some point in the system, it is necessary to have a transition from waveguide to coaxial line or vice versa. There are many ways to join a coaxial line to a waveguide so that there is little loss due to reflection.

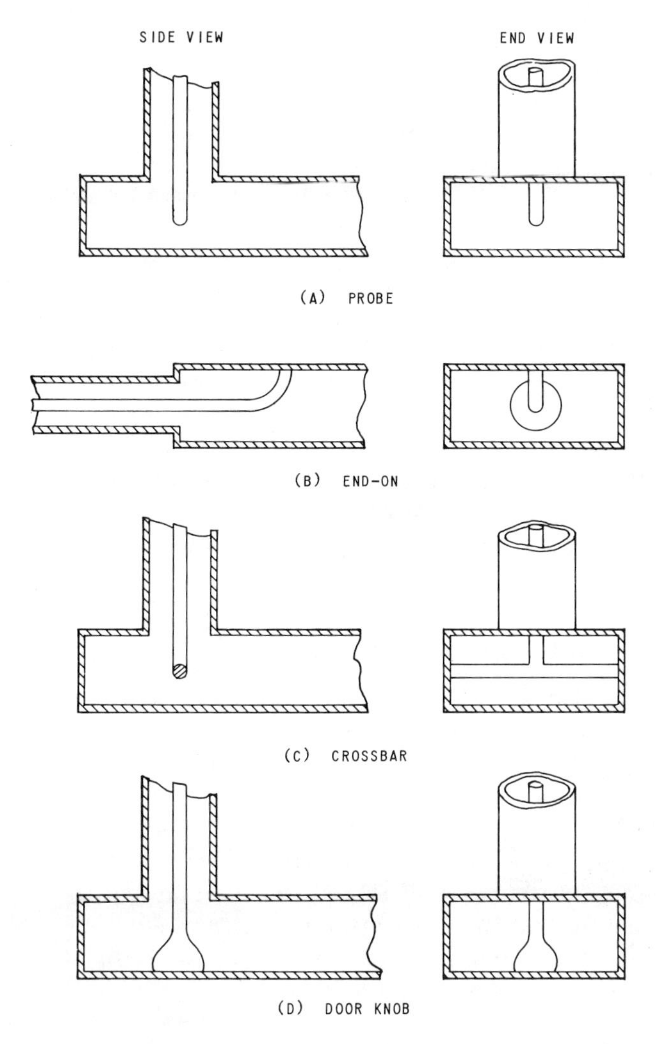

Fig. 8.8. Waveguide-to-coax transitions.

In all of them, the center conductor of the coaxial line extends into the waveguide and some portion of it is parallel to the E-field in the guide.

Figure 8.8 shows a few types of waveguide-to-coax transitions; the simplest (as shown in Fig. 8.8a), has the outer conductor of the coaxial line attached to a broad wall of the waveguide, and the inner conductor extended into the waveguide like a short antenna. One end of the waveguide is shorted. The transition can be matched by varying four parameters: the diameter and depth of penetration of the probe, the distance to the end plate or short circuit, and (occasionally) the distance from the center line of the waveguide. A match of 1.2 over a 10 per cent band is not unusual.

Figure 8.8b shows an end-on or collinear transition. The outer conductor of the coaxial line ends on the end plate, and the inner conductor extends into the waveguide and is bent to be parallel to the voltage lines. The end of the inner conductor is fastened to a broad wall of the waveguide. Parameters which can be varied are the distance between the end plate and the point where the loop is fastened to the waveguide, the angle or curve of the bend in the loop, and, of course, additional matching elements in the waveguides. A match of 1.1 over a 30 per cent band can be achieved.

Figure 8.8c shows a cross-bar transition, and Figure 8.8d shows a door-knob. Both of these require additional matching elements and can be made to have a 1.1 match over a 30 per cent band also.

In practical cases, these transitions are made with a very short section of solid coaxial line attached to the waveguide. The other end of the coaxial line has a fitting which can be connected to a standard cable connector.

8.8. CRYSTAL AND BOLOMETER HOLDERS

A crystal or bolometer in a transmission line must receive the microwave power with very little reflection, since any reflection represents a loss and consequently a potential source of error. In both of these devices the microwave signal is changed to a d-c or low frequency signal which must be measured. It is important that none of the microwave energy leaks into the measuring circuit to produce an error there.

In a waveguide, one end of the crystal is usually grounded and the other end attached to the inner conductor of a coaxial line. In order to keep the microwave signal from entering the measuring apparatus, a large bypass capacitor is used at the junction of the coaxial line and the waveguide. Matching the microwave signal to the crystal is a straightforward procedure. In a coaxial line, the crystal is usually put in series with the inner conductor, and a by-pass capacitor is used on the output end. To have a closed d-c circuit, it is sometimes necessary to add a d-c return on the input side of the crystal. Sometimes in a waveguide system, it is convenient to use a waveguide-to-coax transition and a coaxial crystal holder. If the transition has a shorted inner conductor, as in the cross-bar type, it will serve as a satisfactory d-c return for the coaxial holder.

8.9. ROTARY JOINTS

In radar operation, the antenna usually scans the skies while the rest of the system remains fixed. Thus it is necessary to provide a joint which can allow motion but presents no discontinuity. This means even though there is electrical continuity, there must be no contact. The basic element, then, is a choke joint, similar to the flange choke joint illustrated in Fig. 4.12.

Since the rotary joint must have no discontinuity, even with continuous rotation, the two parts of the choke joint must be round and have field patterns which are circularly symmetric. The coaxial line operating in the TEM mode obviously has those qualifications. There are also two symmetric modes in circular waveguide which can be used, the $TM_{0,1}$ and the $TE_{0,1}$. In practice, rotary joints are either coaxial or use the circular $TM_{0,1}$ mode. In all coaxial systems, of course, the coaxial rotary joint would be used. In a rectangular waveguide system, however, the rectangular guide would be changed to coaxial line or circular guide in the $TM_{0,1}$ mode by the proper transitions. Two identical transitions, one on each side of the rotary joint, are necessary.

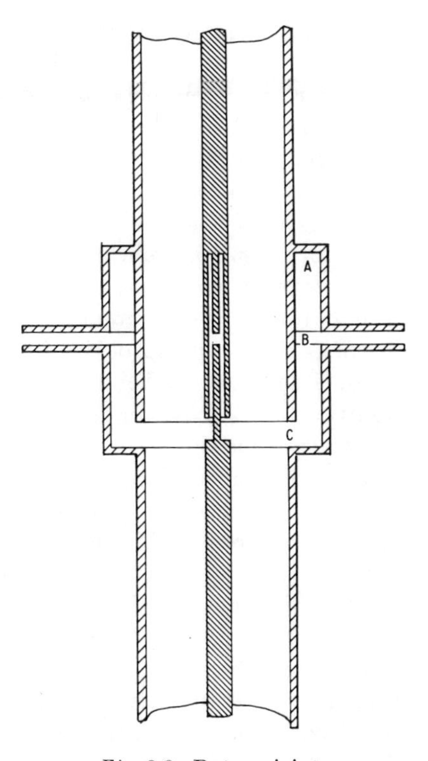

Fig. 8.9. Rotary joint.

Figure 8.9 shows a cross section of one type of coaxial rotary joint. The waveguide joint is similar to the coaxial, but without the inner conductor. In effect, another coaxial line is placed around the main line, and the outer conductor of the main line is the inner conductor of the added coaxial line. At point A, this added line is short-circuited. A quarter-wavelength away, at B, the impedance is transformed to an open-circuit or infinite impedance. The impedance of the juncture at B is in series with this open circuit and must be added to infinity so that the net result is still an open circuit. At C, again a quarter-wavelength away, the open circuit is transformed to a short circuit again, so that there is continuity for the microwave signal. The choke in the inner conductor operates in the same manner.

In practice, suitable bearings are supplied outside of the junction so that the motion is smooth. Also, in order to prevent leakage when slightly off the center frequency, a lossy absorber is placed just outside the joint, at B.

8.10. FERRITE COMPONENTS

By using ferromagnetic material in the transmission line and a suitably oriented magnetic field, it is possible to fabricate components which are nonreciprocal, including a three-port junction which is well-matched in all three ports. Although these components operate on different principles from other microwave elements, the measurement procedures are the same.

A ferrite *isolator* is a nonreciprocal two-port component which will pass microwave signals with low loss in the *forward* direction but will absorb energy in the *reverse* direction. Although different from ordinary attenuators, the measurements required on it are of the same type, but because it is nonreciprocal, measurements must be made in both directions. Thus, the VSWR must be measured in both directions. The low insertion loss in the forward direction, and the high attenuation in the reverse are measured just as for ordinary attenuators. All of these are measured over the frequency band of interest just as for other components. An additional figure of merit used to describe an isolator is its *back-to-front ratio*. This is simply the reverse attenuation in decibels divided by the forward insertion loss in decibels.

A ferrite *circulator* is a multiport component in which energy fed into the first port goes only to the second, energy fed into the second goes only to the third, etc.; energy fed into the last port goes only into the first. Ideally there should be high isolation or attenuation from any port to all ports other than the one next to it. The most common circulators are three-ports and four-ports.

As with isolators, the quantities to be measured are VSWR at all ports, insertion loss in the forward direction (port 1 to port 2, etc.), and isolation between all other ports.

In some ferrite components the magnetic field is furnished by an electromagnet which can be varied, rather than by a fixed permanent magnet. The properties of these components must then be measured at different settings of the magnetic field if the variations are specified. Ferrite components are built in both coaxial lines and waveguides.

8.11. FIXED MISMATCH

In microwave circuits and systems in general, it is necessary to have as low an input VSWR as possible. A primary consideration is, of course, to minimize loss due to reflection. A VSWR of 1.50 will produce a reflection of less than 4 per cent of the power, which is negligible in many applications; thus it would seem unnecessary to try to improve it. However, there are other factors which may necessitate an improved match. For example, a transmitting tube is affected by the mismatch it sees, and even a low mismatch can sometimes cause it to shift operation to a new frequency or a lower output power. A high-gain amplifier must also see a good match, since the reflected power can return to the input and be amplified, causing oscillation.

Tubes, amplifiers, and other elements which can be affected by poor matches are usually tested by putting a specified mismatch at the output and noting the effect. Since the degree of change depends on phase as well as amplitude of the reflected signal, it is usual to have a standard mismatch which can be moved along the transmission line.

The standard mismatch can be a preset probe which is slid in a slotted line and produces a capacitive mismatch. For a small band of frequencies this is adequate, but since the capacitive reactance is a function of frequency, this method will not produce a constant reflection over a large frequency band. Another method which yields a flatter mismatch with frequency uses an output line of different characteristic impedance from the input line. The VSWR is the ratio of the two impedances. A third method uses an attenuator followed by a short circuit. The value of attenuation is chosen so that it absorbs all the reflected power except the amount which the required reflection would produce. Standard mismatches are built with standing-wave ratios from 1.2:1 to 2:1.

QUESTIONS AND PROBLEMS

8.1. It is necessary to have a load with a very small reflection coefficient. Show by means of a block diagram how one can check to see if a load meets the specifications. Comment on the sources of error in your method.

8.2. Explain why a carbon card has more effect in the center of the guide than it does along the side.

8.3. What is the attenuation in db/ft for the $TE_{1,0}$ mode in a copper guide 2 in. by one in. at:
a. 3 Gc
b. 6 Gc *ans.* b. 0.008 db/ft
c. 18 Gc

8.4. A coaxial attenuator has the form as shown in Fig. 8.5. The line is 1 in. in diameter, air filled, with a characteristic impedance of 50 ohms. What is the proper length for the cut-off attenuator section if the total attenuation is to be 50 db at 1 Gc? At 5 Gc?

8.5. A section of copper X-band guide (0.9 in. by 0.4 in.) is to be used as an attenuator at 2 Gc. What is the attenuation of a 10-cm section?

8.6. Describe how a simple phase shifter might be constructed. How could the range of this device be checked (use a block diagram)?

8.7. It is required that two waveguides be joined. One is 0.9 inch by 0.4 inch the other is 0.9 inch by 0.6 inch. The center of the band to be matched is 8.6 Gc. Draw a diagram, with dimensions, of a way this might be done.

8.8. What are the four parameters used when matching a rectangular waveguide section to coax? Explain how each might be used.

8.9. The problem is to feed a rotating antenna. The main line is to be a rectangular guide. Show the diagrams for at least two methods of accomplishing this. Label all modes.

8.10. Explain the operation of a choke joint.

8.11. Give three methods of making a standard mismatch. Which is the best from a frequency standpoint?

8.12. How does a ferrite attenuator differ from a carbon card?

9

RESONANT CAVITIES AND FILTERS

Figure 9.1 shows a typical resonant circuit at lower frequencies. It consists of an inductance, L, a capacitance, C, and a shunt resistance, R, which represents unwanted losses in the circuit. The resonant frequency, f, is then

$$f = \frac{1}{2\pi\sqrt{LC}} \tag{9.1}$$

If a similar resonant circuit is required at microwave frequencies, say at 3000 megacycles, it is apparent that the L and C of Eq. (9.1) must be too small to be practicable. The capacity associated with even the smallest

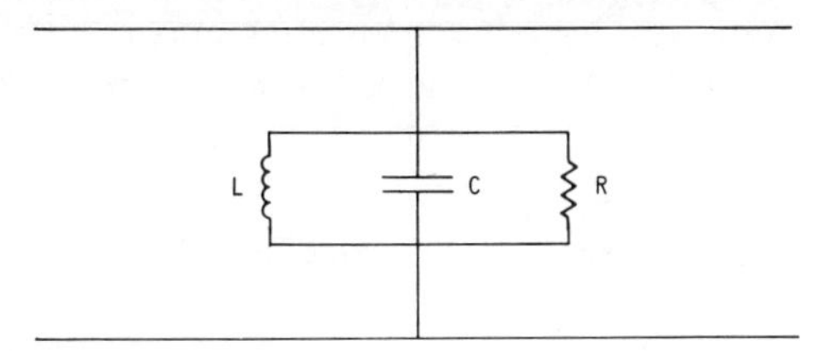

Fig. 9.1. Parallel resonant circuit.

inductor and the inductance of even the shortest leads on the capacitor would upset all calculations. The losses represented by R in Fig. 9.1 would become more appreciable as the frequency was increased. At microwaves, a distributed type of resonant circuit is more practical than the low frequency lumped-element circuit. Filters, too, at low frequencies, use inductances and capacitances which in various combinations form resonant circuits that determine the characteristics of the filter. Again, at microwaves the practice is different: filters use distributed elements rather than lumped-element circuits.

9.1. RESONANT CAVITY

When a wave is propagating in a waveguide, the electric and magnetic fields exist and travel in definite patterns which were illustrated in Figs. 4.2, 4.3, and 4.4. If a short circuit is placed across the waveguide, there is complete reflection and the returning wave exhibits the same patterns. As was explained in Chap. 2, the waves traveling in opposite directions produce a standing wave in the shorted waveguide. At the short, there is a voltage minimum, and this minimum is repeated at half-wavelength intervals from the short circuit. (The wavelength used is, of course, the guide wavelength.) If a short circuit is now placed at one of the voltage minima, there will be complete reflection back toward the first short and in phase with the original signal. This means that the cavity formed by the waveguide and the two shorting plates can support a signal which apparently bounces back and forth between the two plates. The problem of getting the signal into the cavity will be considered later. It is evident, that the length of the cavity must be some integral multiple of half a guide wavelength. These considerations apply to any shape of hollow cavity, but in practice, microwave waveguide cavities are either rectangular or cylindrical.

Since standing waves can also exist in a coaxial line, it should be possible to build a coaxial resonant cavity. Its short-circuiting plates are spaced exactly a half-wavelength apart, since $\lambda_g = \lambda_0$ for coaxial lines which are air-filled.

If the cavity is filled with a dielectric whose relative dielectric constant is ϵ, this value must be used in calculating λ_g in order to determine the resonant length of a waveguide cavity. In coaxial lines the value of λ_g is simply $\lambda_0/\sqrt{\epsilon}$, and the length must be an integral multiple of half of this value.

9.2. MODES IN CAVITIES

Since there are an infinite number of waveguide modes, and since each one can exist in a cavity whose shorting plates are correctly spaced for that mode, it follows that there are an infinite number of cavity modes, also. Just as the waveguide modes are of two types, TE and TM, and have subscripts designating their configurations, the same applies to the cavity modes. The cavity mode is designated by three subscripts.

The first two designate the waveguide mode, and the third indicates the length of the cavity in half guide-wavelengths. Thus a $TM_{2,1,2}$ mode in a rectangular cavity is the $TM_{2,1}$ mode of rectangular waveguide, and the length of the cavity is two half-wavelengths or exactly λ_g. A $TE_{1,1,1}$ mode in a cylindrical cavity indicates the dominant mode in the round guide and a cavity length half of the guide wavelength. In addition, in round waveguide cavities, there are TM modes in which the electric field is everywhere parallel to the axis of the cylinder. The designations for these modes have a zero as the third digit and do not represent possible waveguide modes.

In coaxial line the transmission is usually in the TEM mode. It is customary then to refer to the mode in a coaxial cavity as the coaxial mode and to indicate its length in wavelength; for example, one speaks of "a half-wavelength" coaxial cavity, or, "a two-wavelength" coaxial cavity. However, just as a $TE_{1,1}$ mode and other higher modes may exist on a coaxial line if the frequency is high enough, so also their corresponding cavity modes can exist in a coaxial cavity.

9.3. RESONANT FREQUENCY

As has been indicated, a cavity will be resonant at any frequency for which its length is an integral multiple of half the guide wavelength. This indicates one important difference between the resonant cavity and its low frequency counterpart which was illustrated in Fig. 9.1. The low frequency circuit has *one* resonant frequency. At frequencies higher than the resonant frequency, the circuit appears inductive while at lower frequencies it is capacitive. The resonant cavity, however, has *many* resonant frequencies. It is resonant for every mode and every frequency in which its length is one half-wavelength, two half-wavelengths, or, in fact, any integral multiple of a half. It may be exactly one half-wavelength long for two different frequencies excited in two different modes.

As will be shown later, the coupling mechanism which brings the signal into the cavity must be located in such a way as to excite the desired mode in the cavity. Most probes, loops, and coupling holes can excite more than one mode. Therefore, it is important to know whether a cavity of given dimensions can be resonant in more than one mode at a given frequency or near a given frequency. Now a mode in the waveguide cavity is designated $TE_{m,n,p}$ or $TM_{m,n,p}$ where p represents the number of half-wavelengths. The length ℓ of the cavity is thus

$$\ell = \frac{p\lambda_g}{2} \tag{9.2}$$

The value of λ_g in terms of λ_0 and λ_c is given in Eq. (4.2) This can be substituted in Eq. (9.2) and the new equation is solved for λ_0. The solution gives the resonant wavelength and is designated λ_r.

$$\lambda_r = \frac{1}{\sqrt{(1/\lambda_c)^2 + (p/2\ell)^2}} \tag{9.3}$$

The value of λ_c for any mode may be inserted in Eq. (9.3) to determine the resonant wavelength for that mode. For a rectangular waveguide, substituting the general value of λ_c in Eq. (9.3), the resonant wavelength for the $TE_{m,n,p}$ or $TM_{m,n,p}$ modes is

$$\lambda_r = \frac{2}{\sqrt{(m/a)^2 + (n/b)^2 + (p/\ell)^2}} \tag{9.4}$$

Equation (9.3) applies to cylindrical as well as rectangular waveguides. The letter p is an integer representing half-wavelengths, but, as previously pointed out, it can also be zero for TM modes in round waveguide.

If the rectangular cavity is a cube; that is, if $a = b = \ell$ in Eq. (9.4), then a particular mode can exist in three different directions. There will be many modes which have the same resonant frequency. This is termed a degeneracy. In the case of a cubic cavity there is a twelve-fold degeneracy. If only two sides are equal, there is a four-fold degeneracy. Even if all sides are unequal, there is still a degeneracy in that the TE and TM modes with the same subscripts will have the same resonant frequency. If due to slight errors in manufacture the cavities are not perfectly rectangular, the degenerate modes can be separated slightly. This results in two resonant frequencies which are close together, an obviously undesirable occurrence.

The number of resonant frequencies for a rectangular cavity is, of course, infinite, since the length can be any number of half-wavelengths. However, in practice, there is an upper frequency limit beyond which resonances will have no effect and are therefore of no interest. With a limiting maximum frequency or minimum wavelength, the number of resonances is finite and is given approximately by

$$N = \frac{8\pi V}{3\lambda_m^3} \tag{9.5}$$

where V is the volume of the cavity and λ_m is the minimum wavelength of interest.

The resonant wavelength of a cubic cavity can be found from Eq. (9.4). For the $TE_{1,0,1}$ mode, it turns out to be

$$\lambda_r = \sqrt{2}\, a \tag{9.6}$$

which is the length of the diagonal of a face of the cube.

For cylindrical cavities it is possible to determine the resonant wavelength for specific modes from Eq. (9.3). A few important ones follow; for the $TE_{1,1,p}$ modes,

$$\lambda_r = \frac{2}{\sqrt{(1/0.853d)^2 + (p/\ell)^2}} \tag{9.7}$$

For the $TM_{0,1,p}$ modes,

$$\lambda_r = \frac{2}{\sqrt{(1/0.653d)^2 + (p/\ell)^2}} \tag{9.8}$$

For the $TE_{0,1,p}$ modes,

$$\lambda_r = \frac{2}{\sqrt{(1/0.420d)^2 + (p/\ell)^2}} \tag{9.9}$$

For the $TM_{0,1,0}$ mode in Eq. (9.8), $p = 0$. Then,

$$\lambda_r = 1.306d \tag{9.10}$$

The $TE_{0,1}$ mode in round waveguide is not used except in special applications. However, its corresponding cavity mode is important because of its low attenuation which results in an extremely high Q for the cavity.

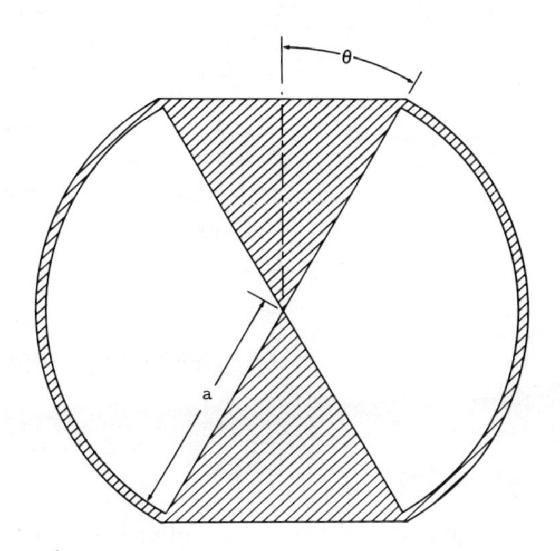

Fig. 9.2. Cross section of spherical cavity with reentrant cones.

Cavities do not have to be derived from waveguides. In fact, resonant frequencies can be found for any hollow enclosure of any shape or configuration—although most odd shapes cannot be analyzed readily. Two which have been analyzed are the spherical cavity and the spherical cavity with reentrant cones. The first resonant wavelength of a sphere of radius a is

$$\lambda_r = 2.28a \tag{9.11}$$

and the second is

$$\lambda_r = 1.4a \tag{9.12}$$

A sphere with reentrant cones is shown in cross section in Fig. 9.2. The resonant wavelength of this type of cavity is

$$\lambda_r = 4a \tag{9.13}$$

and is independent of the cone angle.

9.4. QUARTER-WAVE CAVITY

A quarter-wavelength of line which is short-circuited is equivalent to an open circuit. Therefore, if a signal is reflected by a short circuit,

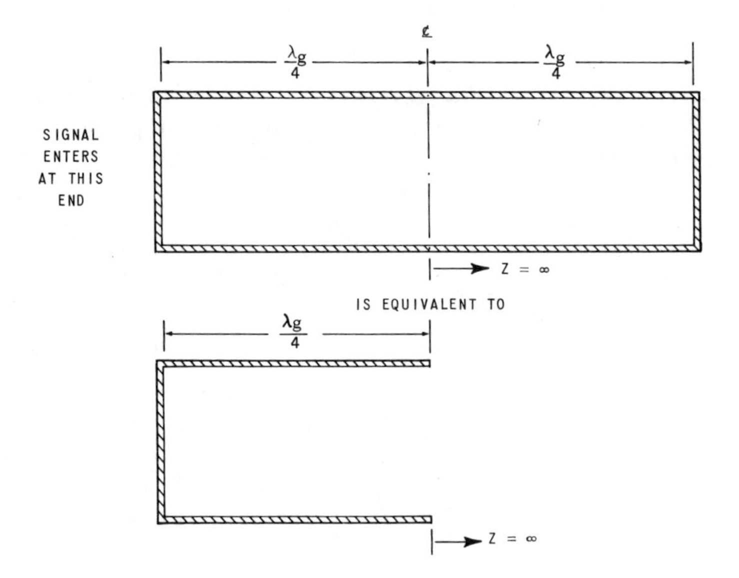

Fig. 9.3. Development of quarter-wave cavity.

it should be subject to the identical reflection by an open circuit which is a quarter of a wavelength nearer. It would seem logical then that a signal introduced at one end of a cavity could be made to resonate, if the other end of the cavity is open-circuited instead of short-circuited and is brought a quarter of a wavelength closer. This is indicated in Fig. 9.3. However, in the case of waveguide cavities this is impossible, because it is impossible to have an open-circuited guide. A waveguide without a termination is actually terminated by free space which has a finite impedance, and, in fact, instead of complete reflection from an open guide there will be some loss due to radiation into free space. It is true that a waveguide below cut-off does appear as an open circuit, and perhaps the cavity could be tapered down to achieve this, but it would probably not be practical or have any advantage over a half-wave cavity.

In a coaxial cavity it is simple to achieve an open circuit. The center conductor is cut; if the outer conductor is below cut-off for any mode at the specified frequency, it will look like a below cut-off waveguide and hence an open circuit at the end of the inner conductor. This is illustrated in Fig. 9.4.

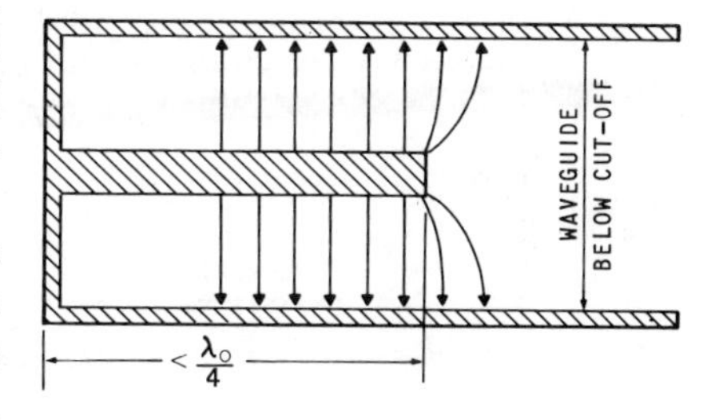

Fig. 9.4. Quarter-wave coaxial cavity.

Because of fringing fields, as shown in Fig. 9.4, the effective length of a quarter-wave cavity is greater than its physical length. Thus, the inner conductor is actually less than a quarter-wavelength in length. However, there is no appreciable saving in space since the outer conductor must be longer than the inner in order to prevent radiation.

A quarter-wave cavity has higher order resonances just as any other cavity, but with one important difference. The *half-wave* coaxial cavity is resonant whenever the length is an integral number of half-wavelengths; that is, it is resonant at all harmonics of its fundamental or first resonance. Similarly, half-wave waveguide cavities are also resonant at frequencies near all harmonics, the shift being caused by a change in the guide wavelength. In the *quarter-wave cavity*, the length must be an odd number of quarter-wavelengths. This means that only odd harmonics will resonate. Thus, the quarter-wave cavity can be used over a three-to-one frequency range without ambiguity whereas other cavities are usable only over an octave.

9.5. COUPLING

Up to now it has been assumed that a resonant cavity is a completely enclosed chamber which favors specific frequencies in the same manner that the low-frequency analog of Fig. 9.1 favors a single frequency. However, it is necessary somehow to introduce the desired signal into the cavity, and this means that there must be a break in the enclosure. As might be expected, the larger the break or the tighter the coupling, the greater will be the effect on the resonant frequency of the cavity.

There are in general three methods of coupling to a cavity: (1) loop coupling, (2) probe coupling, and (3) coupling through an aperture.

These are illustrated in Fig. 9.5. The first two are used to couple coaxial lines to cavities of any type; the third is usually restricted to waveguides coupled to waveguide cavities, although it is sometimes used to couple waveguides to coaxial cavities.

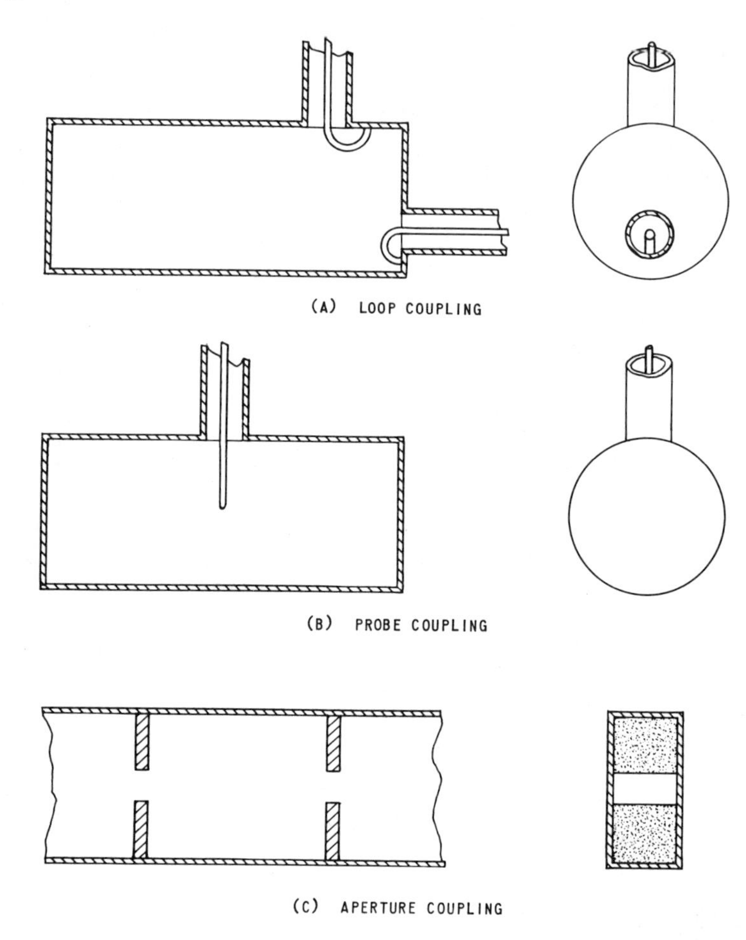

Fig. 9.5. Coupling to a cavity.

If a loop is used to excite a cavity, it must be introduced in such a way that lines of magnetic force of the desired cavity mode can thread through it. This is shown in Fig. 9.5a. Two loops are shown in the figure to indicate that the mode can be excited in more than one location. If the loop is parallel to the lines of magnetic force it will not couple the coaxial line to the cavity. Consequently, it is obvious that the degree of coupling or excitation can be controlled by rotating the loop.

If a probe is used to excite the cavity, it should be inserted at a point of maximum strength in the electric field and must be parallel to this field. As shown in Fig. 9.5b, this point is near the center of a half-wave cavity. In longer cavities a probe is usually coupled one quarter-wavelength from one end. The coupling can be controlled by varying the amount of insertion.

Figure 9.5c shows aperture coupling as used in a waveguide. The aperture may be an iris, a hole, or a slot. It is necessary for the opening to interrupt the flow of current both in the waveguide and in the excited mode in the cavity. As shown in the figure, a simple rectangular cavity can consist of two sets of irises in a waveguide.

If a cavity has just one coupling mechanism which acts as an input to bring power into the cavity, it then behaves as a selective absorber, removing power at the desired frequency and having little effect on other frequencies. A second coupling mechanism may be added as an output. In this case, only the frequency which is accepted at the first coupling will emerge at the output. The first cavity with one exciter is called an *absorption cavity;* the second is a *transmission cavity.*

9.6. Q-FACTOR

The Q or *Q-factor* of a cavity has essentially the same meaning as the Q of a resonant circuit at lower frequencies. At resonance $Q/2\pi f$ is the ratio of the maximum energy stored to the power loss; that is,

$$Q = \omega \frac{\text{Maximum energy stored}}{\text{Power loss}} \tag{9.14}$$

The power loss in the cavity is due to currents in the walls and is thus proportional to the skin depth, δ, given by Eq. (2.1). The energy stored depends on the shape of the cavity. A good approximation for the Q of a cavity is

$$Q \approx \frac{V}{A\delta} \tag{9.15}$$

where V is the volume of the cavity, A is the surface area of the cavity, and δ is the skin depth.

The maximum Q attainable depends on the dimensions and shape of the cavity as well as the material from which it is made. For a rectangular cavity, maximum Q is achieved when the three dimensions are equal; that is, when the cavity is a cube. For a cylindrical cavity, maximum Q occurs when the diameter equals the length. A coaxial cavity has maximum Q when the ratio of its conductors $b/a = 3.6$. This is the ratio for minimum attenuation, as was explained in Sec. 5.4. Although the resonant frequency of a spherical cavity with reentrant cones does not depend on the angle of the cones, the Q does vary with this angle. The maximum Q occurs when the angle θ is 33.5°, where θ is the half-angle shown in Fig. 9.2.

The coupling devices connected to a cavity lower the apparent Q. Thus, in specifying the Q of a cavity, it is necessary to differentiate between the intrinsic or *unloaded* Q of the cavity and the total or *loaded* Q which includes the effect of coupling mechanisms. These are designated

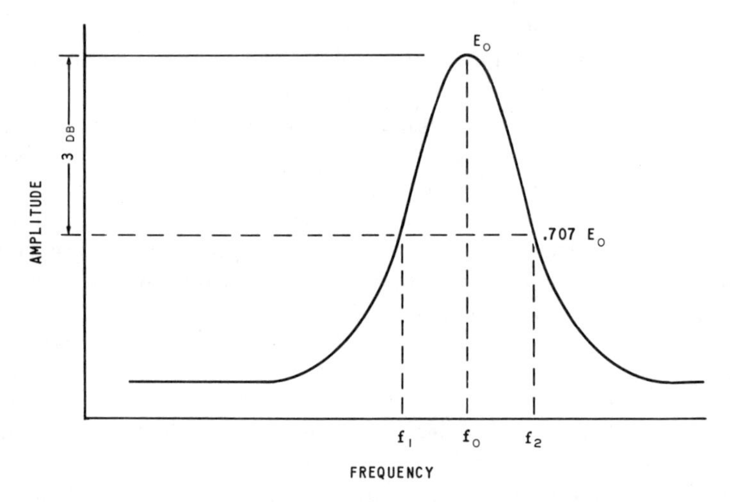

Fig. 9.6. Resonance curve.

Q_U and Q_L, respectively. The Q of the coupler or exciter is designated Q_E (with a subscript if there are more than one coupling device and they differ). The loaded Q is related to the unloaded Q and external Q's as follows:

$$\frac{1}{Q_L} = \frac{1}{Q_U} + \frac{1}{Q_{E_1}} + \frac{1}{Q_{E_2}} \tag{9.16}$$

The loaded Q of a cavity may be measured very simply by observing the shape of its resonance curve as a function of frequency. A typical response curve is shown in Fig. 9.6. Half-power frequencies f_1 and f_2 are measured. The Q_L of the cavity can then be determined from

$$Q_L = \frac{f_0}{f_2 - f_1} \tag{9.17}$$

9.7. CAVITY MEASUREMENTS

The measurements typically made on a resonant cavity include the resonant frequency, the Q, VSWR, insertion loss (of a transmission type), and temperature effects. Measurement of loaded Q has already been discussed. Measurements of VSWR and insertion loss are no different from similar measurements on other components. To measure frequency, a signal is coupled to the cavity, and the frequency of the signal is varied until resonance is observed. This will appear as a dip in the output if an absorption cavity is used or as a peak with a transmission cavity. Then the signal frequency which caused resonance is measured with a calibrated tunable cavity.

When the metal cavity is heated, it expands, and thus its resonant frequency changes. To overcome this, cavities are usually made of materials, such as invar, which have a very low coefficient of thermal expansion. Compensated cavities include mechanisms which maintain a constant internal size even when the parts change with temperature. In order to measure the effects of temperature, the cavity is placed inside a controlled oven, and measurements are made through holes in the walls. A thermocouple is usually attached directly to the cavity so that its temperature can be known during the measurements.

9.8. APPLICATIONS OF CAVITIES

Cavities have many uses in microwave circuits—both as circuit elements and as measuring instruments. Fixed cavities are used as elements in filters and as frequency controls in microwave oscillators. Tunable cavities are used as wavemeters and receiver preselectors. Both types are used as echo boxes.

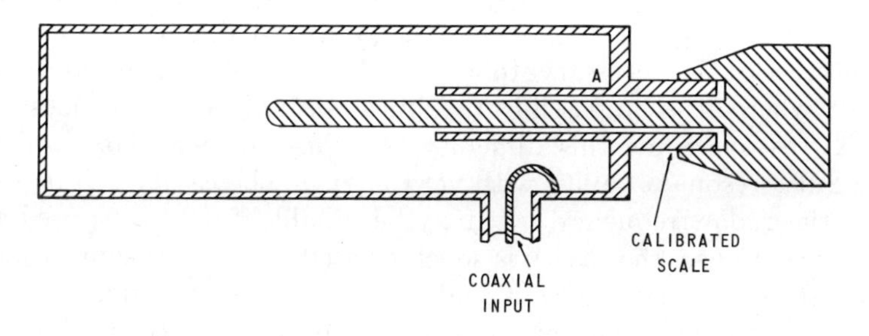

Fig. 9.7. Coaxial wavemeter.

A cavity is made tunable by varying one of its dimensions, usually the length, and calibrating the variation. A typical coaxial wavemeter is shown in Fig. 9.7. In this cavity, the variation of insertion of the inner conductor into the chamber varies the resonant frequency. The center conductor is usually an extension of the plunger of a micrometer so that its motion may be measured. To calibrate this wavemeter, the resonant frequency is determined for different settings of the micrometer, that is, for different insertions of the center conductor. After a calibration chart is made, the cavity itself may be used to measure frequency. It should be noticed that the sliding contact is made near the center of the cavity rather than at the obvious point at the end (point A in Fig. 9.7). Point A is a point of high current in the cavity and would require a good r-f connection so it is usually avoided, although satisfactory wavemeters have been made without the sleeve shown in the figure.

A cylindrical waveguide tunable wavemeter is shown in Fig. 9.8. It should be noticed that as the resonant cavity is made smaller, the back chamber gets larger and may introduce a false resonance. The lossy material on the back of the plunger tends to damp out any resonances in the back chamber.

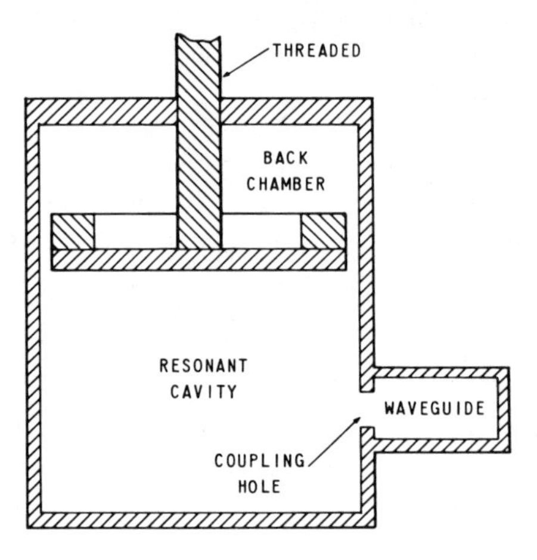

Fig. 9.8. Cylindrical wavemeters.

Tunable cavities may also be used as preselectors in microwave receivers. Here a transmission cavity is required. Since only the resonant frequency can get through, the preselector permits tuning to individual signals. If the preselector is calibrated as a wavemeter, it also indicates the frequency of the input signal.

Sometimes it is necessary to check the over-all performance of a radar set where it is impossible to obtain suitable target echoes. The usual way to overcome this difficulty is to use an echo box, which is nothing but a resonant cavity with very high Q. The cavity can be fixed-tuned to the radar frequency or it may be tunable if it is to be used with many radars. In use, the cavity is shock-excited by the transmitter pulse, and because of the very high Q, it continues to "ring" long after the transmitter pulse is ended. The energy reradiated to the receiver from the echo box is an exponentially decaying signal. The time required for the signal to disappear from the receiver display is an indication of the overall performance. A typical set-up is shown in Fig. 9.9.

In many microwave tubes, the element that determines frequency is a resonant cavity which is built into the tube envelope. The electron beam passes through the portion of the cavity where there is maximum electric field so that there is greatest interaction between the electrons and the r-f energy.

Cavities are sometimes used to measure the dielectric constant and loss tangent of dielectric materials. A small piece of the material to be measured is put into a cavity whose resonant frequency and Q are known. There will be a shift in the resonant frequency due to the dielectric constant of the material and an increase in loss caused by dielectric losses. The increase in loss will result in a reduced Q. The exact relationships depend on the size and shape of the material and its location in the cavity.

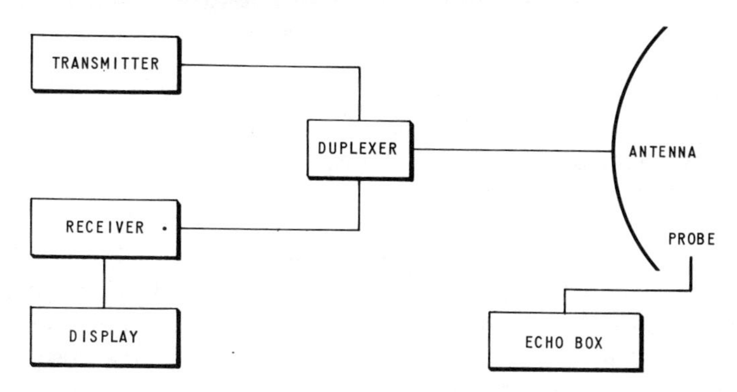

Fig. 9.9. Echo box set-up.

Since a cavity is selective by itself, it can be rightfully assumed that fixed-tuned cavities, coupled together in series, will act as a filter, just as low frequency tuned circuits do. Cavities are used principally in band-pass filters. These and other filters will be discussed in the rest of this chapter.

9.9. FILTERS

A filter is a network which will pass a specified band or bands of frequencies and will stop or reject other frequencies. The frequencies which are transmitted through the network with little or no attenuation are said to be in the *pass-band* of the filter. Those which are rejected are in the *stop-band.*

The rejected frequencies may be reflected back to the input or may be absorbed in resistive loads. In a reflecting filter the input impedance in the stop-band should be a pure reactance, and in an absorbing filter it should be real and equal to the characteristic impedance of the transmission line. In both types, the filter should be well-matched in the pass-band.

There are four general types of filters which are designated according to their characteristics. A filter which passes a band of frequencies and stops frequencies above and below this band is called a *band-pass* filter.

As will be shown in the next section, a band-pass filter can have more than one pass-band, the higher bands being harmonically related to the first. Just as a filter can pass a band of frequencies, so also it can reject a band of frequencies. A filter of this type which rejects a band of frequencies but passes all below and above this band is called a *band-reject* or a *band-stop* filter. If the pass-band of a band-pass filter extends from direct current up to a specified frequency, that is, if the filter passes all frequencies below the specified frequency, it is called a *low-pass* filter. The specified frequency in this case is called the cut-off frequency. Similarly, a *high-pass* filter passes all frequencies above a specified cut-off frequency.

9.10. BAND-PASS FILTER

A simple resonant cavity of the transmission type is a band-pass filter, since it passes a very narrow frequency range and stops (by reflection) all frequencies above and below this range. The higher the Q of the cavity, the narrower will be the pass-band and the steeper will be the "skirts" between the pass-band and the stop-bands. If a wider pass-band is required, the Q can be lowered. If a steep transition is also required between the pass-band and a stop-band, a second cavity (also with the same low Q) can be added in series. A band-pass filter usually consists of several cavities in series; the Q of the individual cavities or sections is made low enough so that the whole pass-band will be transmitted. Then enough sections are used so that the rejection at specified frequencies in the stop-band is high enough. In general, the more sections there are, the greater will be the insertion loss in the pass-band.

The cavities or sections in a band-pass filter may be coupled directly or they may be separated by quarter-wavelength transmission lines. The former case results in a shorter filter; in the latter case, individual sections can be tested and adjusted before assembly.

Since a cavity is resonant at many frequencies, it should be expected that a band-pass filter using cavities will have higher pass-bands. This is usually the case, although if the higher pass-bands are outside the range of interest, it is unimportant. However, if no higher pass-band can be tolerated, it is necessary to use in series two filters that have different higher pass-bands.

9.11. LOW-PASS FILTER

At lower frequencies, a low-pass filter is constructed from lumped circuit elements and may resemble Fig. 9.10. The same prototype can be used at microwaves by substituting a high impedance line for an inductance and a low impedance line for a capacitance. In a coaxial line, this is done simply by abruptly changing the diameter of the inner conductor; in strip line also, this is accomplished by changing the width of the strip. As with a band-pass filter, a low-pass filter at microwaves

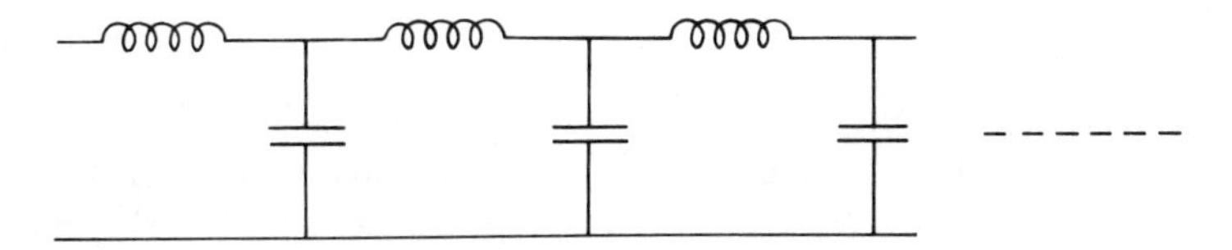

Fig. 9.10. Low-pass filter prototype.

exhibits a pass-band at some higher frequency harmonically related to the cut-off frequency.

9.12. HIGH-PASS FILTER

Since a waveguide has a cut-off frequency below which there is no transmission, the waveguide by its very nature is a high-pass filter. However, the skirt or transition from pass-band to stop-band is too gradual for most applications. High-pass filters with steep skirts are difficult to construct. However, satisfactory results for most applications can be achieved by using a band-pass filter in which the lower cut-off frequency is the required cut-off of the high-pass filter. The pass-band of the filter is made broad enough to include all the frequencies of interest. The upper cut-off is then unimportant.

9.13. BAND-STOP FILTER

A band-stop filter is usually a combination of a low-pass and a band-pass filter. The cut-off frequency of the low-pass filter is selected as the lower cut-off of the stop-band. Similarly, the lower cut-off of the band-pass part of the filter is the upper cut-off of the required stop-band. The pass-band of the band-pass section is made high enough to include all frequencies of interest.

9.14. FILTER MEASUREMENTS

The description of a filter's characteristics or desired characteristics is an indication of the measurements to be performed. The filter should have low VSWR and insertion loss in its pass-bands. These are usually specified as maximum values which must not be exceeded. Since measurements should be checked at every frequency in the pass-band, it is usually desirable to use a sweep-frequency technique to determine VSWR values. The frequencies at which the insertion loss is three decibels are frequently specified and easily verified. Stop-band specifications usually include minimum values of attenuation at specified frequencies. The VSWR in the stop-band is also specified for absorbing filters, but is not mentioned for reflecting filters.

Measurements are straightforward but certain precautions must be noted. When measuring what should be a high attenuation (in a stop-band) it is important to make sure that no power is reaching the detector through a higher pass-band. For example, suppose a low-pass filter has a cut-off of 2000 megacycles and exhibits a higher pass-band at 6000 megacycles. If the frequencies of interest do not exceed 4000 megacycles, the higher pass-band is not important. However, it can result in an erroneous reading when the filter is checked. The specifications may require 60 decibels of attenuation at 3000 megacycles. When this is checked with the signal generator set at 3000 megacycles, a second harmonic from the generator can get through the pass-band at 6000 megacycles and apparently show a low attenuation at 3000 megacycles. This particular problem can be solved by using a low-pass filter on the signal generator. If this filter has a cut-off slightly above 4000 megacycles, the 3000 megacycle signal from the generator will pass through it and reach the filter under test, but the 6000 megacycle harmonic will not be transmitted.

9.15. APPLICATIONS OF FILTERS

Filters are used in measurement set-ups wherever it is likely that an unwanted spurious signal or harmonic will produce an erroneous result. The preceding paragraph illustrated a typical situation where a harmonic had to be "filtered out." Filters are also used to observe a particular portion of a signal, when the signal is not monochromatic. By using a very narrow-band, high-Q cavity as a tunable filter, it is possible to pass selective portions of an oscillator output and thus obtain a clearer picture of the output signal.

In a microwave superheterodyne receiver, a narrow band-pass filter can be used to pass the desired signal and reject the image. A band-pass filter is sometimes used between the local oscillator and the mixer in order to prevent spurious signals and harmonics in the local oscillator from reaching the mixer.

In some microwave systems, one antenna is connected to two receivers, each of which is to receive a different frequency. To accomplish this, it is necessary to use an arrangement of filters which receives all frequencies in one input and then separates them in two outputs in a prescribed way. This arrangement is called a *diplexer*. It consists of an input connected to two outputs with a filter in each output line. The filter in a particular line will accept the frequency specified for that line and reflect the unwanted frequency. The position of the filter in the line is adjusted so that the reflected signal from the filter adds in phase in the other line.

More filters and more lines can be added to separate input signals into three, four, or more channels according to frequency. The resultant arrangements are named for the number of channels they create; hence the arrangements immediately following the diplexer, are the triplexer and quadruplexer.

QUESTIONS AND PROBLEMS

9.1. Find the lowest resonant frequencies of the two cavities shown to the right. What is the proper nomenclature for each of these modes?

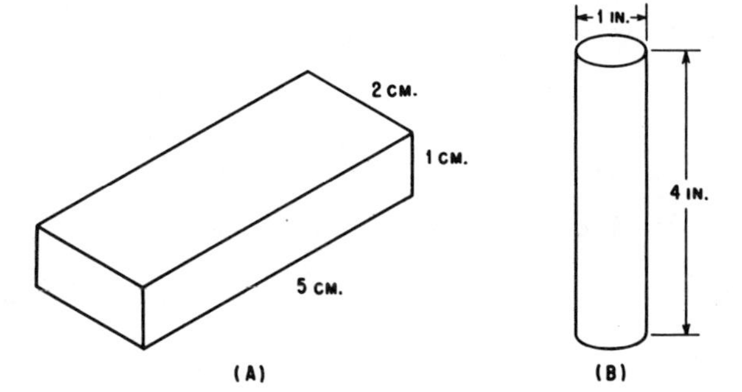

Fig. P 9.1.

9.2. What is the maximum number of resonances below 12.4 Gc for the rectangular cavity of Prob. 9.1?

9.3. What is the resonant frequency of a cube 2 cm on each side? *ans.* 10.6 Gc

9.4. Illustrate three methods of coupling energy into a cavity.

9.5. You have a cubic cavity 10 cm on each side. Find
a. The lowest resonant frequency. *ans.* 2.13 Gc
b. The Q, if the walls are made of copper. *ans.* 23,000
c. What is the effective shunt resistance? *ans.* 7.8×10^6 ohms

9.6. Repeat Prob. 9.5 for a cylindrical cavity similar to the one shown in diagram b of Prob. 9.1.

9.7. What is the resonant frequency of the cube of Prob. 9.3 if the cavity is completely filled with a substance with a dielectric constant of 2.6?

9.8. What Q is required for a cavity if the bandwidth is to be 10 per cent at 10 Gc? At 1 Gc?

9.9. Distinguish between loaded and unloaded Q.

9.10. A copper cavity has a Q of 12,000. The inside is now plated (the depth is greater than skin depth) with silver. What is the new Q?

9.11. Show with diagrams how loops and probes can be used to couple energy into a rectangular cavity operating in the $TE_{1,0,1}$ mode. Draw the magnetic and electric fields in the diagrams.

9.12. Draw a block diagram for a set-up to check the insertion loss of a filter over a band of frequencies.

9.13. Draw output voltage versus frequency curves for the following types of filters: band-pass, band-stop, high-pass, and low-pass.

9.14. Draw the low frequency equivalent circuit for each of the filters of Prob. 9.13.

9.15. Show how the definition $Q = X/R$ where X is the magnitude of the resonant reactance of either the coil or the capacitor and R, the series circuit resistance, is equivalent to

$$Q = \frac{2\pi(\text{energy stored})}{\text{energy dissipated/cycle}}$$

9.16. Draw a block diagram and explain how you could measure the Q of a cavity.

9.17. How could a cavity be used to check the dielectric constant of a dielectric?

10

MIXERS AND DETECTORS

In the early days of radio, receivers usually used a crystal detector. Once the vacuum tube was invented, crystal sets virtually disappeared except as experimenters' curiosities. At microwave frequencies, however, ordinary radio tubes were unsatisfactory. The distances between electrodes were appreciable parts of wavelengths, and the inter-electrode capacitances were undesirable "lumped" elements in the circuit. Consequently, when microwave receivers were designed for radars, the crystal diode was again used as the detector of the receiver. It was an improved, better-packaged diode with its cat-whisker completely enclosed, but it was still the same type of crystal diode that was used in the first receivers.

10.1. PHYSICAL DESIGN

There are two general types of structure used for crystals, a ceramic cartridge and a coaxial type. The ceramic cartridge type is shown in Fig. 10.1. The sensitive element is a *silicon* wafer. A thin wire called a cat-whisker makes contact with this wafer. The rest of the cartridge is simply a uniform holder to protect the crystal element. A ceramic body surrounds the silicon and cat-whisker, and the metallic contacts are threaded into the ceramic. The larger contact is called the *base* and in most applications is usually grounded. In Fig. 10.1, the silicon wafer is attached to the base, and the cat-whisker to the contact prong. This arrangement may be reversed, in which case the crystal is said to have *reversed polarity*. A double-ended cartridge is also available; it uses a contact prong at each end; a separate base is supplied which may be attached to either prong. This type of cartridge may be used for either forward or reverse polarity. In all cartridge types the metal parts are gold plated for resistance to corrosion.

The coaxial type of crystal is simply a coaxial line, whose characteristic impedance is 65 ohms, terminated in the same arrangement of silicon wafer and cat-whisker. This structure is indicated in Fig. 10.2. The center

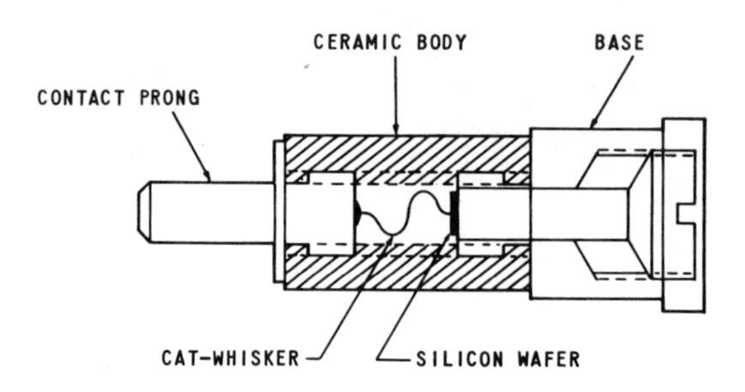

Fig. 10.1. Ceramic cartridge diode.

conductor or *pin* is supported by an insulating bead coaxially with the outer shell. The shell is grounded in most applications. As with the cartridge crystal, the polarity of the coaxial type may also be reversed.

In Sec. 8.8, it was indicated that a by-pass capacitance was usually built into a crystal mount in order to keep the r-f signal out of the metering circuit. A coaxial crystal with built-in bypass capacitance is frequently used to simplify the crystal holder. A view of this *tripolar* crystal is shown in Fig. 10.3. The end-plug (which is normally grounded in the coaxial

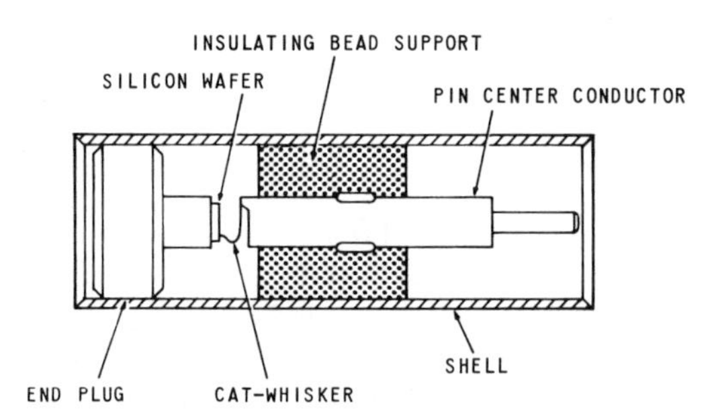

Fig. 10.2. Coaxial diode.

type) is insulated from the outer shell in the tripolar device. The capacity between the plug and the shell is about five to ten picofarads, which is sufficient to by-pass the microwave signal. The back terminal can be connected directly to an IF or video amplifier as required with no additional by-pass capacitance needed in the crystal holder.

The tripolar crystal can be used over a broad frequency range, covering several octaves. For frequency bands less than an octave, the other two are generally used, the cartridge up to about 10,000 megacycles, and the coaxial type above this frequency.

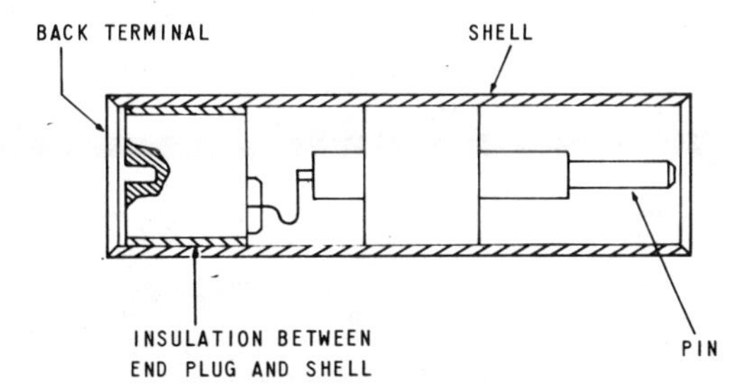

Fig. 10.3. Tripolar diode.

The over-all length of a cartridge type crystal is about 0.82 inch. The coaxial and tripolar crystals are about three-quarters of an inch long. On all crystals, the diameters are less than a quarter-inch.

10.2. EQUIVALENT CIRCUIT

The equivalent circuit of a crystal diode is shown in Fig. 10.4. Schematically the circuit is the same for coaxial and cartridge types and for mixers and detectors. The barrier capacitance, C, is about 0.1 or 0.2 picofarads, and the spreading resistance, r_s, is about 20 ohms. However, the value of R_b, the barrier resistance, depends on the crystal application and the direction of current.

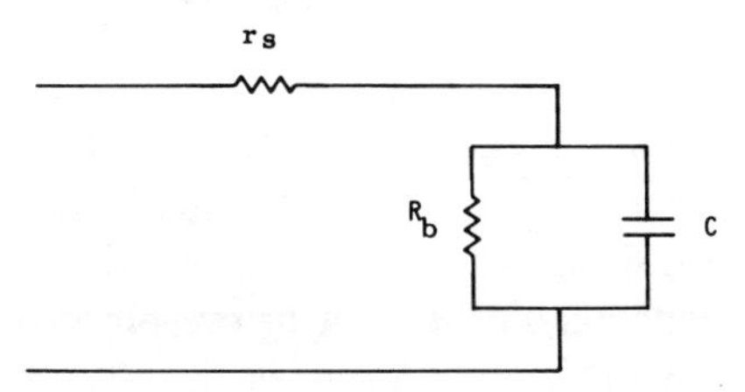

Fig. 10.4. Diode equivalent circuit.

A typical d-c current-voltage curve for a 1N21 crystal is shown in Fig. 10.5. In the forward direction, the response "obeys the square law"; that is, the current varies as the square of the voltage. Actually this is true only for small incident powers up to about 15 microwatts (below the knee in Fig. 10.5). Above this value, the response is linear. The value of R_b at the low-power level is about 10,000 ohms in either direction. At the comparatively high power level, where the crystal is linear, the value of R_b drops to about 200 ohms in the forward direction but is unchanged in the reverse direction. The *back-to-front ratio* of resistances here should be at least 20:1 if the crystal is undamaged.

At low levels, the crystal may be used as a sensitive, direct detector of microwave signals. The crystal may be operated without any d-c bias or it may have a small d-c bias to improve sensitivity. As a mixer, the

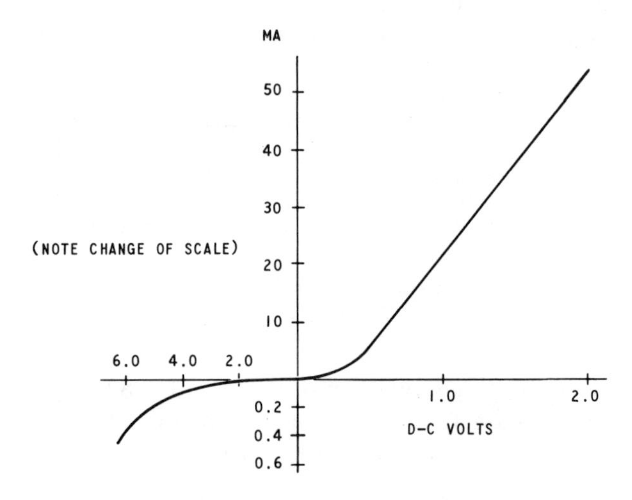

Fig. 10.5. Typical 1N21 d-c characteristics.

crystal is usually operated in the linear portion at a power level of one-half to one milliwatt. It is "biased" to this point on the curve by a local oscillator.

10.3. DETECTOR CRYSTALS

In the simplest form of microwave receiver, the incoming signal is demodulated directly by the crystal. The video components of the signal (corresponding to the modulation envelope) are then fed into a video amplifier and other electronic circuits; the selection of these circuits depends on the type of presentation desired. Used in this way, the crystal is called a *crystal-video detector* and the receiver is called a *crystal-video receiver*. A crystal-video receiver is much less sensitive than a superheterodyne receiver where the crystal is used as a mixer, but its extreme simplicity makes it useful in many applications where size, weight, and complexity are important considerations.

The r-f bandwidth of a crystal-video receiver is usually limited by the antenna or other components ahead of the crystal, rather than by the crystal itself. Receivers of this type have been built covering the frequency range from below 100 megacycles to more than 10,000 megacycles, a range of more than 100:1. As long as the r-f signal can get into the crystal, it will be detected. The limiting factor then is VSWR. But even if the VSWR is five to one, there is less than a three decibel loss in sensitivity. From Eq. (2.25), the voltage reflection coefficient from a VSWR of five is $\frac{2}{3}$. The power reflected is the square of this coefficient, $\frac{4}{9}$. Thus, $\frac{5}{9}$ of the power gets to the crystal to be detected.

10.4. VIDEO IMPEDANCE OF DETECTORS

The video impedance of the diode is simply the d-c resistance of the diode in the microwatt region. As was indicated in Sec. 10.2, this resistance is about 10,000 ohms and is the same in both the forward and reverse directions; it can be changed by applying bias to the crystal. It is measured by applying a d-c voltage of a few millivolts, noting the current on a microammeter, and calculating the resistance from Ohm's Law.

Although the video impedance can vary from crystal to crystal, there are both upper and lower limits on its value. If the resistance is too high, poor fidelity of signal response will result. This is evidenced by distortion of the pulse shape, although it may also produce an increase in sensitivity. If the pulse shape is not important, values of resistance up to 100,000 ohms may be used. A low value of video impedance, less than 2000 ohms, produces a reduction in sensitivity.

10.5. D-C BIAS

The term, "bias," applied to a video crystal always means d-c bias in the forward direction. (Negative bias always causes a degradation in performance and consequently is never considered.) The crystal is biased by applying a voltage across it so that a current of about 25 microamperes is produced. The best value of this current will vary with the crystal and with the particular application, but it usually will be between 10 and 50 microamperes.

The video impedance and the r-f impedance both vary with bias. Consequently, a change in bias produces changes in the input match, in sensitivity, and in pulse fidelity. If pulse fidelity is the important consideration, the bias is adjusted for best reproduction of the input pulse shapes. This will usually decrease the sensitivity slightly, but occasionally it actually improves sensitivity. If pulse shape is not important, the bias is adjusted for maximum sensitivity.

10.6. SENSITIVITY OF DETECTORS

The sensitivity of a video crystal is a measure of its ability to receive weak signals. The more sensitive the receiver, the weaker the signal that can be received. The sensitivity of the receiver is simply the strength of the weakest signal which will produce a specified output. There are two different output specifications used: *minimum detectable signal* (MDS), and *tangential sensitivity* (TS). Both are expressed in *decibels below a milliwatt* or −dbm. Thus a signal strength of one microwatt is 30 decibels below a milliwatt or simply −30 dbm.

The minimum detectable signal or MDS is, as its name implies, the strength of the weakest signal which can be detected. Unfortunately this is not a reliable specification, since to a large extent it depends on the experience of the operator and the type of presentation (visual or aural). The ability to pick out a signal from the background noise improves with experience.

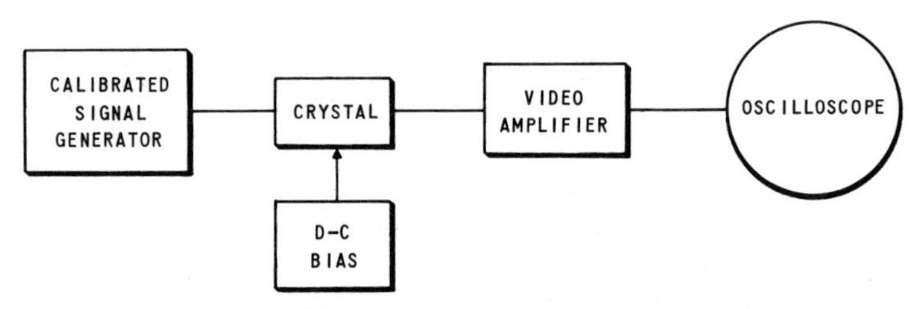

Fig. 10.6. Set-up for measuring sensitivity.

Tangential sensitivity is a more reliable measure, since different operators will obtain consistent measurements of this quantity. The term applies to the power of a signal which is just equal to the background noise; that is, the signal-to-noise ratio is unity. Figure 10.6 illustrates a typical set-up for measuring tangential sensitivity. The pulsed output from the signal generator is applied to the crystal. The output from the crystal is presented on the oscilloscope after one or more stages of amplification. A d-c bias source is included to observe the effect of bias.

The types of presentations seen on the oscilloscope during sensitivity measurements are shown in Fig. 10.7. Incidentally, the same kinds of displays are observed when the receiver and oscilloscope are used to receive microwave signals in the field. In Fig. 10.7a, only the background noise is presented. This is the noise present in the receiver or crystal before any signal is applied from the signal generator. Figure 10.7b shows the proper setting for tangential sensitivity. The signal is raised so that it just equals the noise level. This is accomplished by slowly increasing the output of the calibrated signal generator. The strength of the signal can then be read from the calibration and is expressed in −dbm. It should be noted that as the signal level is increased from zero, a section of the noise presentation moves up on the scope. The signal level is the bottom of this section. When it is tangent to the top of the rest of the presentation, as in Fig. 10.7b, the correct setting has been reached. Figure 10.7c shows the result with a signal which is too weak and Fig. 10.7d indicates a signal which is too strong. Figure 10.7e shows a signal just beginning. This might be MDS for some operators, while others may not notice it.

The minimum signal which can be detected is obviously weaker than the value determined as the tangential sensitivity of the receiver. As a rule of thumb, it is generally assumed that an average operator can spot a signal about four decibels weaker than the value of the tangential sensitivity.

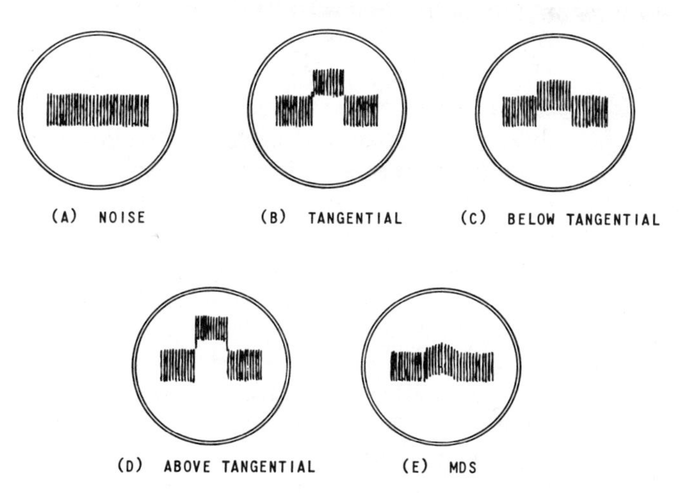

Fig. 10.7. Sensitivity displays.

The sensitivity of a video detector is dependent on the bandwidth of the video amplifier as well as the characteristics of the crystals. The narrower the bandwidth, the more sensitive will be the detector—sensitivity varying inversely with the square root of the bandwidth; that is, if the bandwidth is increased by a factor of four, the sensitivity is decreased by three dbm. Thus, a crystal which has a sensitivity of -55 dbm with a one megacycle bandwidth amplifier will have a sensitivity of -52 dbm with a four-megacycle amplifier. At first glance it would seem desirable to make the bandwidth of the amplifier as narrow as possible in order to increase the sensitivity. However, the video amplifier must be wide enough to pass the frequencies in the pulse envelope. For good pulse reproduction, the bandwidth of the video amplifier should be of the order of the reciprocal of the pulse width and preferably should be twice this value. Thus, to reproduce a one-microsecond pulse ($\delta = 10^{-6}$) the bandwidth should be two megacycles ($2/10^{-6}$). Typical sensitivities achieved with video crystals range from −50 dbm to −60 dbm with a two-megacycle amplifier.

10.7. MIXERS

When a crystal is used as a mixer, it is operated on the linear portion of the curve shown in Fig. 10.5. The operation is exactly like that of a superheterodyne circuit at lower frequencies. A local oscillator furnishes the bias to fix the point of operation on the linear portion. This requires approximately half a milliwatt of local oscillator power for each crystal used in the mixer, although as much as a milliwatt per crystal will be satisfactory. The mixer may consist of a single crystal or a balanced pair. The balanced mixer will be discussed in a later section.

The local oscillator power is usually furnished by a low power klystron, a backward wave oscillator, or other microwave tube. It may also be supplied from a VHF (very high frequency) transistor oscillator followed by suitable harmonic multipliers to bring the frequency into the microwave range. Typically, the local oscillator puts out about ten milliwatts. This is then fed to the crystal through a variable attenuator, to bring the power at the crystal down to the desired half milliwatt. The attenuator then also acts as a pad to isolate the local oscillator from possible variations in the rest of the circuit.

The incoming microwave signal mixes with the local oscillator signal in the crystal, and sum and difference frequencies are produced. The intermediate frequency amplifier is tuned to a fixed frequency which is the difference between the local oscillator frequency and a signal frequency. Common values for the intermediate frequency (IF) are 30 megacycles and 60 megacycles, although the exact value in a particular receiver can be completely arbitrary.

It is obvious that the received signal must be separated from the local oscillator (LO) frequency by an amount equal to the IF, but it may be above or below the LO frequency; that is,

$$f_{\mathrm{RF}} = f_{\mathrm{LO}} + f_{\mathrm{IF}}$$

or (10.1)

$$f_{\mathrm{RF}} = f_{\mathrm{LO}} - f_{\mathrm{IF}}$$

Thus, the receiver can receive two different microwave frequencies, one above and one below the LO frequency, for each setting of the local oscillator. The desired frequency is the signal, and the other is called the *image*. Image rejection at microwaves is accomplished by methods similar to those used at lower frequencies. A tunable filter called a preselector may be used between the antenna and the mixer. The preselector must be tuned synchronously with the local oscillator, but the two are separated in frequency by the intermediate frequency. This set-up has another advantage in that the preselector may be made narrow enough to keep stray local oscillator signals from reaching the antenna and radiating.

Another method of image rejection makes use of the fact that the two i-f signals produced by microwave signals on either side of the local oscillator are 180° out of phase. By a proper phasing circuit, the image can be *out-phased* or rejected while the desired signal is unaffected.

10.8. MIXER SENSITIVITY AND NOISE

Mixers are more complicated than video detectors. A mixer requires a local oscillator and associated power supply and also requires more sophisticated circuitry. The justification is the increased sensitivity. A mixer is about 50 decibels more sensitive (100,000 times) than a video detector.

The sensitivity of a microwave mixer, or, in fact, of any receiver, is limited by the noise generated in the circuit itself. This is power caused by thermal agitation in the input impedance to the receiver, and since it covers all frequencies, it is termed *noise.* The wider the bandwidth of the receiver, the more noise will come in, and consequently the receiver will be less sensitive. In the case of detectors operating in the square-law region of the crystal, it was mentioned in Sec. 10.6 that the sensitivity is inversely proportional to the square root of the bandwidth. With mixers, however, the operation is linear, and the sensitivity is inversely proportional to the bandwidth.

The sensitivity of a mixer is that signal power which is three decibels more than the noise power. This is

$$S_{\mathrm{DB}} = -171 \text{ dbm} + 10 \log B + F_{\mathrm{DB}} \qquad (10.2)$$

where S_{DB} is the sensitivity in $-$dbm, B is the bandwidth in cycles per second, and F_{DB} is the noise figure of the receiver in decibels. Thus, if the receiver has a noise figure of 20 decibels and an i-f bandwidth of two megacycles, $10 \log B = 63$ and,

$$S_{\mathrm{DB}} = -171 + 63 + 20 = -88 \text{ dbm}$$

The term *noise figure* needs some explanation since, from Eq. (10.2), it is definitely the limiting factor on sensitivity when the bandwidth is specified. Noise figure is the ratio of (1) the available signal-to-noise at the input of a network to (2) the available signal-to-noise at the output. The word, "available," means the power that could be transferred in a well-matched system. Expressed as an equation:

$$F = \frac{S_i/N_i}{S_o/N_o} = \frac{S_i}{S_o} \cdot \frac{N_o}{N_i} \qquad (10.3)$$

where F is the noise figure expressed as a ratio, S is signal power, N is noise power, and the subscripts i and o refer to input and output, respectively. The ratio F is sometimes referred to as *noise factor*, and the term F_{DB} is reserved for noise figure. Thus,

$$F_{\text{DB}} = 10 \log F \tag{10.4}$$

Which means that if the noise factor is 100, the noise figure is 20 decibels.

If the receiver were perfect, the signal-to-noise at the input would equal the signal-to-noise at the output, and thus $F_{\text{DB}} = 0$. However, all receivers and networks generate some internal noise. This appears at the output as an increase in N_o of Eq. (10.3) and consequently an increase in F. In order to standardize input noise, N_i is defined as the thermal noise power available from a resistor, at a temperature of 290°K (62.6°F). This is necessary because the noise power is directly proportional to the temperature in degrees Kelvin.

The noise figure, F_{DB} in Eq. (10.2), is the receiver noise figure and is affected by both the diode and the i-f amplifier following it. The diode has some thermal noise which is a function of its physical and chemical characteristics. The ratio of this thermal noise to that of a perfect resistor is designated N_r. The output power of the mixer at IF is less than the microwave input power; the loss, called *conversion loss*, is designated L_c and is measured in decibels. The i-f amplifier itself is not perfect in that it, too, has noise. Its noise factor is designated $N_{\text{i-f}}$. The value of the noise figure taking the statements above into consideration is

$$F_{\text{DB}} = L_c + 10 \log (N_{\text{i-f}} + N_r - 1) \tag{10.5}$$

where F_{DB} is the value for Eq. (10.2), and the other quantities are as defined above. For example, if L_c is five decibels, N_{IF} is four, and N_r is five then

$$F_{\text{DB}} = 5 + 10 \log 8 = 14 \text{ db}$$

Thus far it has been assumed that the transmission line and microwave components between the antenna and the receiver are lossless. Naturally, any loss before the mixer is a corresponding reduction of sensitivity. For convenience, this loss is usually considered part of the noise figure and should be added to the right-hand side of Eq. (10.5). In the example above, if the r-f insertion loss from the antenna to the mixer is 0.7 decibels, the noise figure, F_{DB}, is 14.7 decibels instead of 14 decibels.

10.9. EFFECT OF R-F AMPLIFIER ON NOISE FIGURE

If an r-f amplifier is used between the antenna and the receiver to boost the incoming signal, it would seem that this should increase the sensitivity of the receiver. However, this is not necessarily so. The amplifier itself has a noise factor greater than unity so that effectively it has noise power produced at its input. This noise power is amplified with the signal. After being amplified, the signal must be greater than the noise of the receiver plus the amplified noise of the amplifier.

If the amplifier has a low noise figure and a high gain, it may be combined with a receiver to improve the sensitivity. The sensitivity is still determined from Eq. (10.2), but if the over-all noise figure is reduced, the sensitivity will be improved. The over-all noise factor is given by

$$F_T = F_1 + \frac{F_2 - 1}{G_1} \tag{10.6}$$

where F_T is the total or over-all noise factor, F_1 is the noise factor of the first stage (or amplifier in this case), F_2 is the noise factor of the receiver, and G_1 is the gain of the amplifier. For example, assume a receiver has a noise figure of 14 decibels; its noise factor is 25. Thus $F_2 = 25$. Assume the r-f amplifier has a gain of 13 decibels and a noise figure of three decibels; that is $G_1 = 20$ and $F_1 = 2$. Then,

$$F_T = 2 + \tfrac{24}{20} = 3.2$$

This is a noise figure of about five decibels, or about nine decibels better than without the amplifier. The sensitivity, from Eq. (10.2), will also be improved by nine dbm.

How much improvement will result if a second stage of amplification identical with the first is introduced? Here it can be assumed that the receiver and the amplifier nearest it form one box whose characteristics are known. That is, F_T from the example above is now F_2 in the new case. So the new over-all noise factor is

$$F_T = 2 + 2.2/20 = 2.11 = 3.2 \text{ db}$$

This is an improvement of only 1.8 decibels and probably not worth the additional cost and complexity.

10.10. NOISE FIGURE MEASUREMENTS

If the bandwidth of the receiver is accurately known, its noise figure can be measured by simply using a calibrated signal generator in the set-up shown in Fig. 10.8. The output power meter is set at some convenient point with the signal generator *off*. Thus, the meter is reading only noise. Now the c-w signal generator is turned on and the signal level increased slowly until the output is increased by three decibels. The generator calibration now reads the sensitivity of the receiver. This figure and the known bandwidth are inserted in Eq. (10.2), which can now be solved for the noise figure.

A more reliable method of measuring the noise figure utilizes a noise generator and is illustrated in Fig. 10.9. At microwaves, the noise generator is a gas discharge tube in a suitable waveguide or coaxial line. The output of the tube is constant across the frequency band and is known

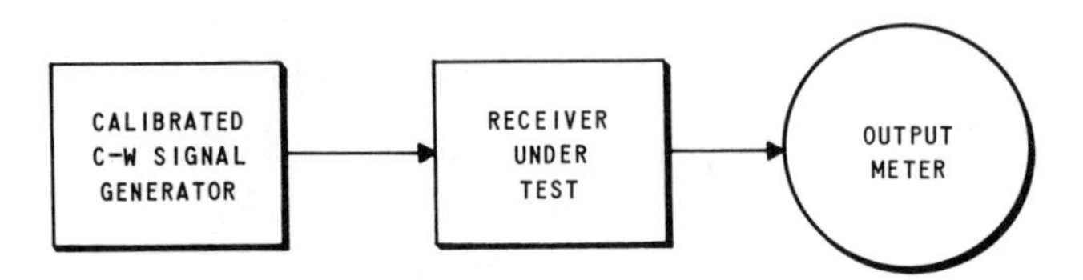

Fig. 10.8. Noise figure measurement using signal generator.

accurately. A load is put at one end to absorb power going in the wrong direction. The noise figure meter at the output of the receiver under test has circuitry which square-wave modulates the noise source. Then the noise output of the receiver with the noise source on is compared to the output with it off. The noise figure is read directly on the meter. This method is faster and more accurate than the signal generator method for low noise figures and does not require a prior knowledge of bandwidth. For noise figures above 20 decibels, however, the noise generator method requires additional circuitry.

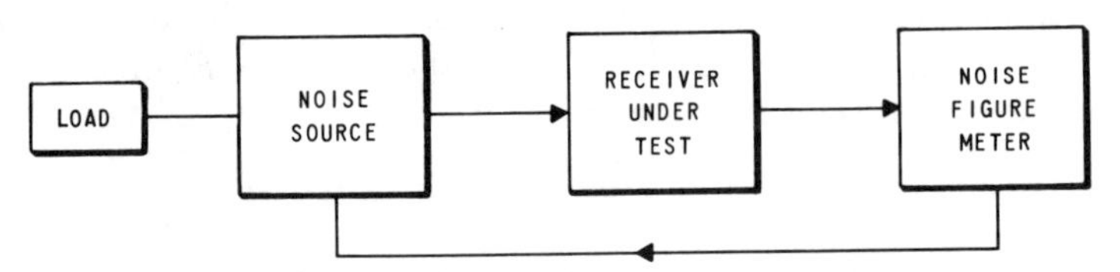

Fig. 10.9. Noise figure measurement using noise source.

In measuring the noise figure of a superheterodyne receiver with a noise source, it is important to know if the receiver has image rejection. With image rejection the noise figure is correct as measured and is termed the single-channel noise figure. However, if there is no image rejection, twice as much noise will pass through the i-f amplifier; that is, the noise source puts out all frequencies, and the frequencies above and below the LO frequency which differ by the IF will go through. Thus, the noise figure read on the meter, called the double-channel noise figure, will be three decibels less than the single-channel value. To get the single-channel noise figure then, it is only necessary to add three decibels to the reading. This problem doesn't arise when the noise figure is measured with a signal generator.

10.11. BALANCED MIXERS

Part of the noise in the receiver originates in the local oscillator. If this noise could be eliminated, there would be a reduction in the over-all noise figure and thus an increase in sensitivity. It is possible to "balance out" local oscillator noise products by using a *balanced mixer*—two types of balanced mixers are shown schematically in Fig. 10.10.

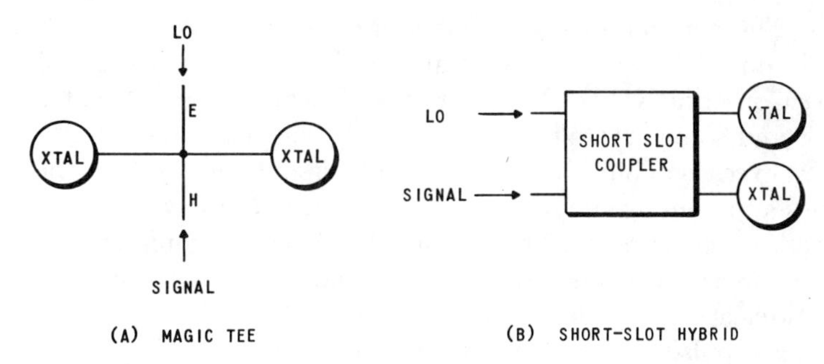

Fig. 10.10. Balanced mixers.

A magic tee balanced mixer is illustrated in Fig. 10.10a. The signal is fed into the E-arm and the local oscillator into the H-arm, or vice versa. Because of the properties of the magic tee, the power fed into the H-arm divides equally, and half arrives at each crystal. The two halves are *in phase*. The power fed into the E-arm also splits equally, and the two halves arrive at the crystals *out of phase*. The phase of the i-f signal produced at each crystal depends on the phases of the two signals which mixed to produce it. The two IF signals will thus be 180° out of phase. These are added in push-pull at the input of the i-f amplifier. Noise in the local oscillator arrives at each crystal in the same phase as the local oscillator signal since both travel the same paths. When this noise beats with the LO signal to produce noise products at the intermediate frequency, the noise products at the two crystals are in phase with one another. These are cancelled when the signals are combined in push-pull.

Figure 10.10b illustrates a short-slot hybrid balanced mixer. The local oscillator and the signal are fed into two adjacent arms, and the crystals are in the other two. Since there is a 90° phase shift through the short-slot coupler, the two halves of the LO power arrive at the crystals

90° out of phase. The two halves of the signal are also 90° out of phase, but the half that lags for the LO leads for the signal. Thus, the two i-f signals are 180° out of phase and can be combined in push-pull. As with the magic tee balanced mixer, the noise products from LO noise are in phase at the two crystals and will be cancelled by the push-pull addition.

It is difficult to manufacture perfectly balanced push-pull transformers. Consequently, the more usual practice is to produce push-pull combination by reversing the polarity of one of the crystals and then combining the two signals in parallel.

10.12. POWER CONSIDERATIONS

The maximum power that a crystal can absorb without damage depends on whether the power arrives in spikes of energy, pulses, or as c-w. Conservatively, 10 or 15 watts of peak power will not damage a crystal, and some crystals will withstand twice this amount. If the peak power is excessive, but not sufficient to burn out the crystals, the damage becomes apparent as an increase in noise figure. For c-w or average power, a conservative figure is 100 milliwatts, but, again, some crystals can be used to three times this value without damage. Manufacturers' recommendations should be followed in all cases.

A sharp pulse of energy produces a local heating effect, but the spike is over before the whole crystal gets heated. For short pulses, one erg of energy is a normal maximum; however, the only unfailing rule to follow in this kind of measurement is to respect the manufacturer's recommendations.

Crystals can also be damaged from voltage breakdown. For safety, three volts of direct current should not be exceeded across the crystal, in either direction. For short pulses, reverse voltage can be as much as ten volts.

10.13. OTHER DIODES

Although not designed primarily for mixing or detection, other semiconductor diodes have special uses at microwaves and should be mentioned briefly in this chapter. Of especial interest are varactors and tunnel diodes.

The *varactor* or *varactor diode* is a special semiconductor which exhibits a capacitance that can be varied by varying the applied voltage. If the applied voltage is a microwave signal, the capacitance will vary at the microwave frequency. The varactor is used in a microwave circuit called a *parametric amplifier*. This type of amplifier has high gain and low noise figure (typically less than three decibels) so that it can be used in front of a receiver to improve its sensitivity.

A *tunnel diode* has a voltage-current characteristic which has a negative resistance region; that means that over part of the diode's range, the current decreases when the voltage is increased. When operated in this range, the tunnel diode acts like a negative resistance amplifier.

Measurements on the tunnel diode amplifier and the parametric amplifier are similar. Gain is measured exactly the same way as attenuation is measured on passive networks, except that when there is positive gain, the strong and weak signals are reversed from their positions in attenuation. Noise figure is measured as indicated in Sec. 10.10. Both gain and noise figure are measured as functions of frequency.

Tunnel diodes can also be used as low power microwave oscillators, as mixers, and as detectors. As a mixer, a tunnel diode requires only about one microwatt of LO power which can be furnished by another tunnel diode used as an oscillator.

Varactors are used as harmonic multipliers in order to obtain microwave signals from UHF and VHF oscillators. The varactor can supply high harmonics, such as the eighth or tenth, with less loss than ordinary crystals. For example a varactor can produce the eighth harmonic with less than a ten decibel loss.

QUESTIONS AND PROBLEMS

10.1. Below what power level can one expect the crystal output to follow the "square law"?

10.2. Draw the equivalent circuit for a tripolar crystal. Label the equivalent components with typical values.

10.3. Define the term "front-to-back ratio" as applied to a crystal. What is a satisfactory value for this ratio?

10.4. Why is d-c bias sometimes applied to crystals in microwave circuits? Discuss this for both detectors and mixers.

10.5. Define: video impedance, detector sensitivity (both MDS and tangential).

10.6. The VSWR in a guide feeding a detector is 4.
  a. What is the power loss due to this mismatch?
  b. What is the reflection coefficient?
  c. What is the return loss?

10.7. Describe in detail, using a block diagram, how to measure the tangential sensitivity of a video detector.

10.8. How is the bandwidth of the video amplifier related to the tangential sensitivity of the crystal detector in a receiver? What practical factors limit the narrowness of the band that can be used in the video amplifier? What bandwidth is necessary to receive radar pulses of 5 μsecs duration?

10.9. What is meant by the term "image frequency"? What means are taken to discriminate against the image frequency?

10.10. Define noise figure and noise factor.

10.11. The noise figure for a receiver is 4 db. The bandwidth is 2 Mc. What is the sensitivity of this receiver? Explain the significance of this number.

10.12. What are the factors that determine whether the addition of an r-f stage will improve the sensitivity of a receiver?

10.13. A receiver has a noise figure of 6 db. A decision is made to place an additional r-f stage in the set. This amplifier has a gain of 15 db and a noise figure of 3 db. What is the gain in the over-all noise figure?

10.14. Draw a block diagram and explain how one can measure the noise figure of receiver.

10.15. What is a "balanced mixer"? Why is it used?

11

SWITCHING

In microwave circuits, as at lower frequencies, it is sometimes necessary to switch energy on and off or to switch it from one path to another. Or again, it may be necessary to send signals of one frequency into one line and of another frequency into a second line. In its broadest sense, any circuit in which there are different paths or states of transmission are switching circuits. However, the word *switch* is reserved for components in which a physical change occurs during the switching operation. A *mechanical* switch is one in which there is a movement of some portion of the component, as for example, a shutter or post in a waveguide. If the movement is accomplished by hand, the switch is a *manually operated* switch; if by a solenoid or other electric means, it is *electromechanical*. An *electrical* switch is one in which a change in electrical characteristics produces the switching action. Diodes which behave as capacitors are typical of the devices used in electrical switches. A change in the voltage across the diode changes the impedance and, thus, the VSWR in the circuit. Switching is accomplished without any mechanical movement. Other switching circuits are duplexers and diplexers. These may or may not contain actual switches.

11.1. MECHANICAL SWITCHES

A simple mechanical microwave switch is a shutter in a waveguide. This is illustrated in Fig. 11.1a. The shutter can be rotated so that it is either vertical, blocking the guide, or horizontal, in which case it is parallel to the broad walls of the waveguide. When it is vertical, it is, of course, in the *off* position and acts as a short circuit (or nearly so because some energy may leak around the edges). When it is horizontal, it behaves as a septum in the waveguide. The power divides on each side of it and recombines after it, so that there is very little reflection or loss. The switch can be rotated manually or by an electric motor.

Another single-pole-single-throw waveguide switch is shown in Fig. 11.1b. Here a post is dropped into the waveguide. With the post out of the circuit, the signal passes without loss or reflection. When the post drops in, it introduces a susceptance which causes reflection and prevents all

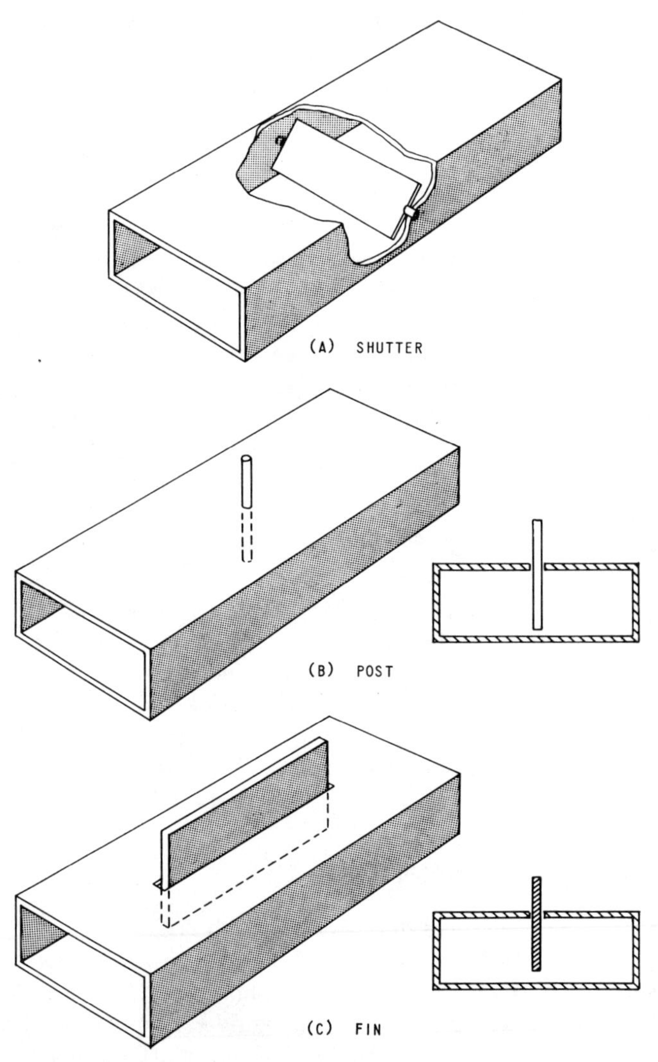

Fig. 11.1. Mechanical switches in waveguide.

the power from reaching the load. The post may be dropped all the way across the waveguide as shown, in which case it is a high inductive susceptance, or it may be dropped partially into the guide to be capacitive. At a critical point it will have infinite susceptance and cause complete reflection as was mentioned in Sec. 4.8.

Another switch in this class consists of a metal fin dropped into a slotted line as indicated in Fig. 11.1c. Here the fin divides the waveguide into two below-cutoff waveguides so that there is complete reflection. With the fin out, the signal passes without loss. Both this type and the post type mentioned above may be operated manually or by means of a solenoid.

If the metal fin of Fig. 11.1c is replaced by a piece of absorbing material, the component is still a switch. With the absorbing card out, there is no attenuation, but when the card is in, it absorbs power and prevents it from reaching the load. It should be noted that in the other switches described above, the power is reflected when it is not passed to the load. The VSWR is high. When an absorbing card is used, however, the VSWR can remain low, and the unwanted power is absorbed instead of reflected. In a sense then, this is a double-throw switch; the power is either sent to its proper destination (in the *on* position) or is thrown into a different load (in the *off* state).

A single-pole-double-throw waveguide switch is shown in Fig. 11.2. The switching portion is a short waveguide elbow in a solid cylinder of metal. This is rotated so that the elbow can be lined up with different guides. In Fig. 11.2a, the input port is connected to output port No. 1. When the cylinder is rotated 90° counter-clockwise, as in Fig. 11.2b, the input port is connected to output port No. 2.

The principle of Fig. 11.2 can be used also for double-pole-double-throw switches, transfer switches, and other special configurations. It is only necessary to provide a suitable number of bends in the rotatable cylinder and properly spaced output waveguides.

In coaxial line, similar mechanical switches can be fabricated. A simple single-pole-single-throw (spst) switch is made by providing a means to open the inner conductor. Usually this is done at a right-angle bend in the coaxial line. In waveguide, the shutter presents a short-circuit, whereas the open inner conductor in coaxial line presents an open circuit, but both produce high VSWR's which prevent power from reaching the load.

A switch similar to the rotating waveguide bend is also made in coaxial line or strip line. A short section of inner conductor is rotated to make contact with the inner conductors of the ports which are to be connected.

The quantities measured on mechanical switches are VSWR, insertion loss, and isolation. In the *on* position, VSWR and insertion loss are

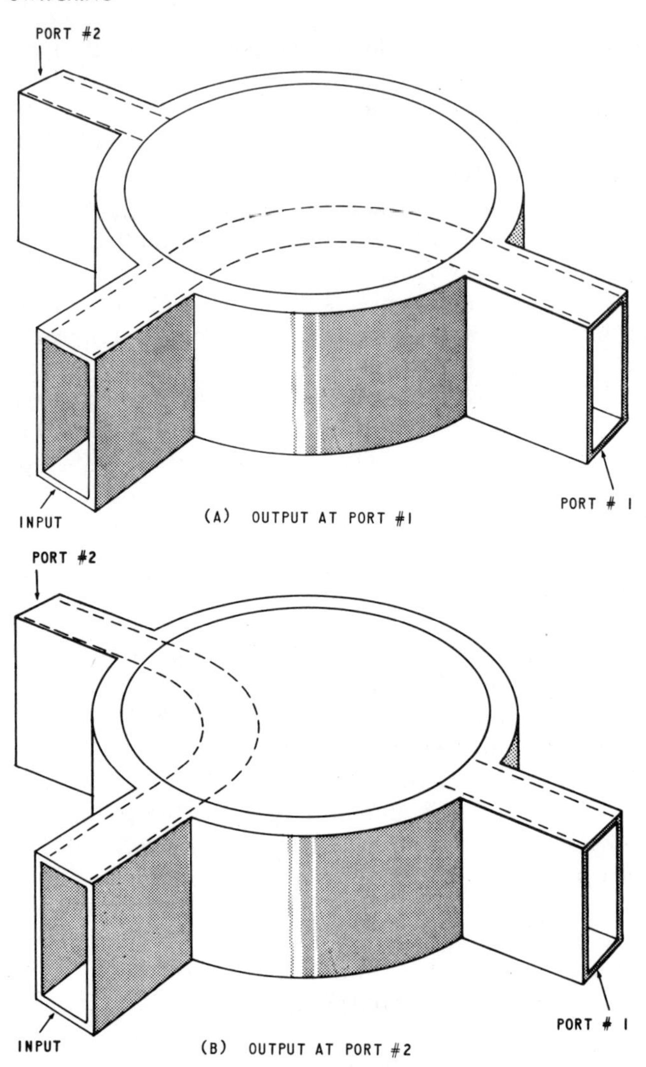

Fig. 11.2. Single-pole-double-throw waveguide switch.

measured by standard methods for all types of switches. In multipole or multithrow types, isolation between the input and the unconnected ports is also measured. Ideally there should be no power coming out of an unconnected port, but there is usually some leakage. Isolation is expressed in decibels. Thus, if the power out of an unconnected port is down 60 decibels from the input power, the isolation is 60 decibels. In single-pole-single-throw switches, isolation is measured in the *off* position. It is the leakage to the load expressed in decibels below input power.

11.2. DIODE SWITCHES

The impedance of a diode depends upon the voltage across it, and microwave switches make use of this fact. The diode is made to have an

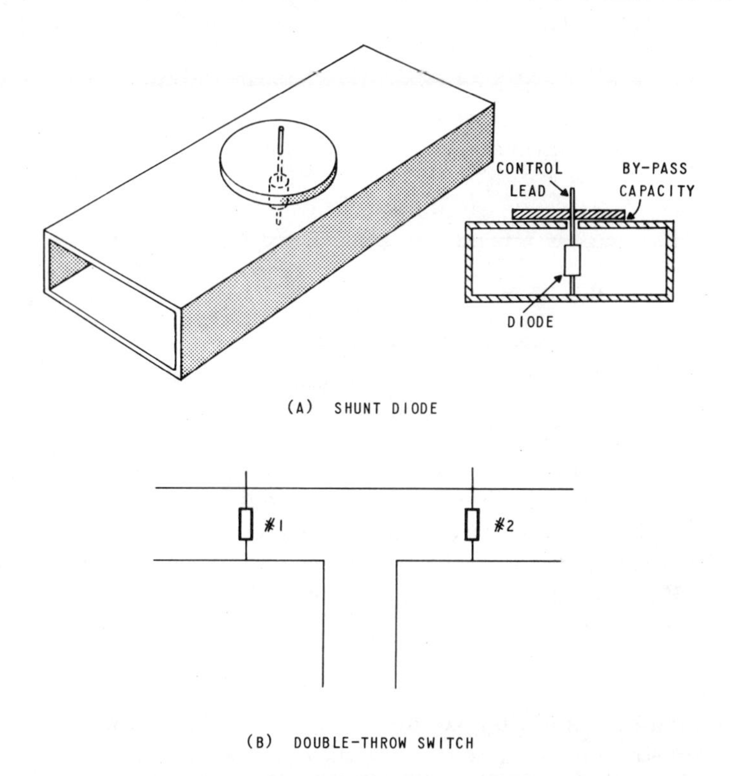

Fig. 11.3. Waveguide diode switches.

impedance which causes no reflection, so that power passes by it and goes to the load. Then the diode impedance is changed by applying a voltage, so that the diode presents a mismatch which reflects most of the power. More diodes can be used to improve the isolation in the *off* condition, but the insertion loss in the *on* condition will also be increased.

Figure 11.3 shows waveguide versions of diode switches. In Fig. 11.3a, the diode is shunted across the waveguide. One end of the diode is connected directly to the guide, but the other end is bypassed to the waveguide through a large capacitor. This furnishes a lead for applying the d-c switching voltage and at the same time presents a short circuit to the guide for the microwave signal. In the *on* position, the diode is biased to have an infinite impedance. This is an open circuit in shunt which has no effect on the signal. Matching reactances may be added to tune out the reactance of the diode leads and package. To change to an *off* condition, the voltage is changed to make the diode reflect the signal.

Two switches of this type may be combined in a tee, as in Fig. 11.3b to make a single-pole-double-throw switch. One diode is biased to reflect and the other to pass the microwave signal. The signal will enter the desired arm and be reflected from the diode in the unwanted arm. The diodes can be positioned so that the reflected signal augments the wanted signal in the desired arm and cancels any junction reflection at the input.

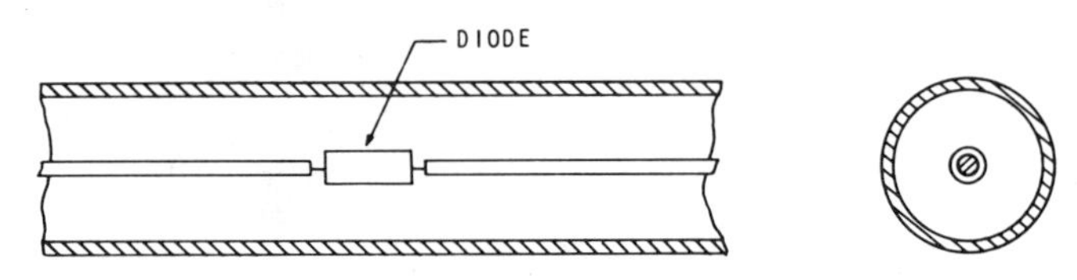

Fig. 11.4. Coaxial diode switch.

In a coaxial line or strip line, the diode is put in series with the inner conductor, as shown in Fig. 11.4. The conditions for *off* and *on* are reversed from those in a waveguide. When the diode in the coaxial line looks like a short circuit, it has no effect on the signal. This is the *on* position. When the diode looks like an open circuit, it will reflect power, and is therefore in the *off* position. As with waveguides, two coaxial diode switches can be combined to make a single-pole-double-throw configuration.

Any kind of diode will give some switching action. However, ordinary detector crystals are not used because they have appreciable d-c resistance which introduces loss in the *on* condition. Special computer diodes are generally used for microwave switches because of their fast switching time, low insertion loss, and large impedance change with small voltage change.

11.3. FERRITE SWITCHES

Ferrite components were discussed briefly in Sec. 8.10. These are nonreciprocal by virtue of an applied magnetic field. Ordinarily in isolators and circulators, the magnetic field is furnished by permanent magnets which are usually outside the waveguide or coaxial line containing the ferrite material. If the direction of the magnetic field is reversed, the losses inside the component are also reversed; that is, low insertion loss directions now have high attenuation, and vice versa. Obviously, if the applied magnetic field is furnished by an electromagnet instead of a fixed magnet, the component will be a switch whose action is controlled by reversing the direction of the field.

11.4. GAS SWITCHES

If a section of a transmission line is filled with gas, there is little effect on microwave signals propagating in the line. In fact, in most cases the dielectric medium is air, which is, of course, a mixture of gases.

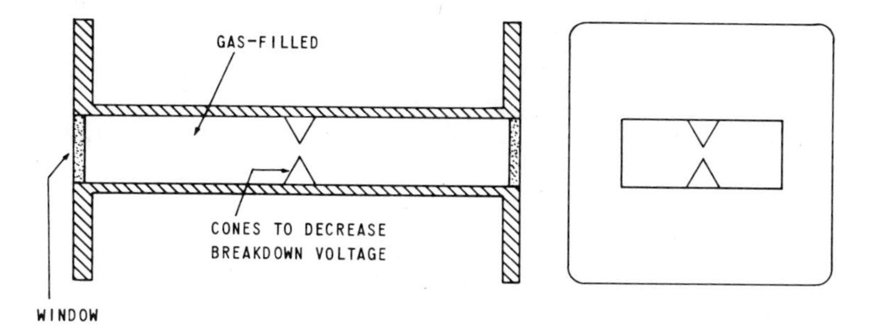

Fig. 11.5. Gas switch.

Replacing the air with a single gas has negligible effect on the propagating signal, unless the signal is so strong that it causes breakdown. When the gas is ionized, as when breakdown occurs, it presents a short circuit to the signal and causes almost complete reflection. This principle is utilized in gas switches.

A simple gas switch consists of a section of waveguide sealed at both ends with windows which are transparent to microwaves. The windows may be glass or ceramic or other material which has low loss and can withstand the high temperatures of the ionized gas. The sealed section is filled with a gas at a very low pressure so that it will ionize easily. Inside the switch or tube is a pair of cones, as shown in the cross-section of Fig. 11.5, which are spaced close together so that breakdown occurs at a lower peak power. This switch is self-triggering. When a pulse of energy arrives at the switch, breakdown occurs if and only if the peak power exceeds a minimum value determined by the spacing of the points and the pressure of the gas. At levels higher than this threshold, the switch breaks down and looks like a short circuit. At lower levels, nothing happens to interfere with the signal, and it propagates with only a little insertion loss. The switch is thus power selective.

If, instead of waiting for self-triggering, the gas tube is fired by a d-c discharge, it can be designed to withstand without breakdown the highest peak powers to which it will be subjected; that is, the gap can be sufficiently large or the gas pressure sufficiently high so that the expected incident power will not cause breakdown. Now if a d-c discharge is fired through the gas tube, it will present a short circuit to all signals, weak as well as strong. It is even possible to discriminate between strong and weak signals by applying across the gap a d-c voltage which is not quite high enough to cause breakdown. Then when an additional voltage, as from an incoming signal, is applied, the tube breaks down.

A third type of gas switch makes use of electron cyclotron resonance in a gas. If a magnetic field is applied across a waveguide perpendicular to the voltage in a guide, it is possible to cause breakdown at levels of the order of milliwatts, while without the magnetic field no breakdown would occur even with kilowatts of power. However, this is frequency selective. The frequency at which breakdown occurs is directly related to the applied magnetic field as follows:

$$f = 2.8H \tag{11.1}$$

where f is the frequency in megacycles, and H is the applied field in gausses. Thus, if the applied field is one kilogauss, breakdown would occur at 2800 megacycles. But signals far removed from 2800 megacycles, for example 4000 megacycles, would *not* cause breakdown. It should be noted, however, that if the gas is ionized because a signal near 2800 megacycles is exciting it, no signal can pass through it.

The electron-cyclotron resonance gas switch is self-triggering as long as the magnetic field is applied; that is, like the first gas switch described above, there is no switching action unless there is a signal present to start it. Once ionization takes place, the signal dominates and, in general, ionization continues until the signal ceases—even if the magnetic field is removed first. The magnetic field controls the switch only in starting the switching action.

Some energy must be removed from the signal to allow a switch to be self-triggering; that is, to allow the incident signal to start and maintain the discharge. This removal results in less than complete reflection. The energy lost in starting and maintaining the discharge is called *arc loss.* In gas switches where the discharge is started by applying a d-c voltage across the gas and is maintained by the applied d-c current, there is very little arc loss.

11.5. MEASUREMENTS

The measurements made on electrical switches include all those made on mechanical switches such as insertion loss and VSWR in the *on* condition and isolation in the *off* condition. These are measured by the methods which have been described. Arc loss of gas switches can be measured by the set-up shown in Fig. 11.6. The signal from a high-power source is fed through a reverse directional coupler to a short circuit. The reflected power is detected at the output of the directional coupler. Then the short circuit is replaced by the gas switch which is fired by the high-power signal. The reflected power at the output of the coupler is compared to the output using the short circuit and is expressed in decibels. Typical

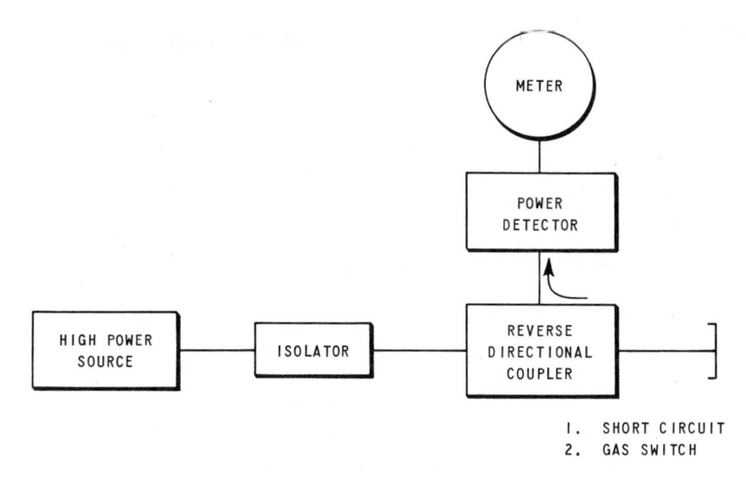

Fig. 11.6. Arc loss measurement.

values are 0.5 to 1.0 decibels for a self-triggering switch and 0.1 to 0.3 decibels with a d-c activated switch.

An important consideration in electrical switches is the time required to switch from one state to another. In mechanical switches, one millisecond (a thousandth of a second) is considered fast. In electrical switches, one microsecond (a millionth of a second) is considered slow. Times quoted for some diode switches are less than five nanoseconds. A nanosecond is a thousandth of a millionth of a second.

The firing time of a gas tube which is triggered by the high power can be measured by the set-up shown in Fig. 11.7. The trace on the oscilloscope is synchronized with the pulse repetition rate of the magnetron. Part of the transmitted pulse travels directly to the detector and the rest travels down the long transmission line to the gas tube. The gas tube is fired by the pulse which is then reflected and appears on the oscilloscope as a second pulse. The horizontal axis of the oscilloscope is calibrated in microseconds. The spacing between the two pulses on the face of the oscilloscope represents the time for the original pulse to travel down the long transmission line, be reflected, and return. If the gas tube is now replaced by a short circuit, the time between the two pulses will be shorter since the reflection at the short circuit occurs immediately. The difference in times with the short circuit and with the gas tube is the firing time of the tube.

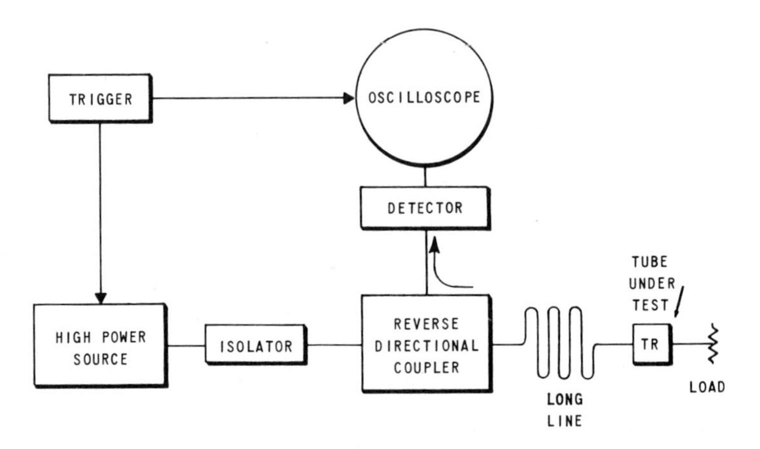

Fig. 11.7. Firing time measurement.

The switching time of switches which are electrically controlled can be measured by the set-up shown in Fig. 11.8. The c-w signal generator feeds a signal at the specified frequency to the switch, which is followed by a suitable detector and oscilloscope presentation. A voltage pulse

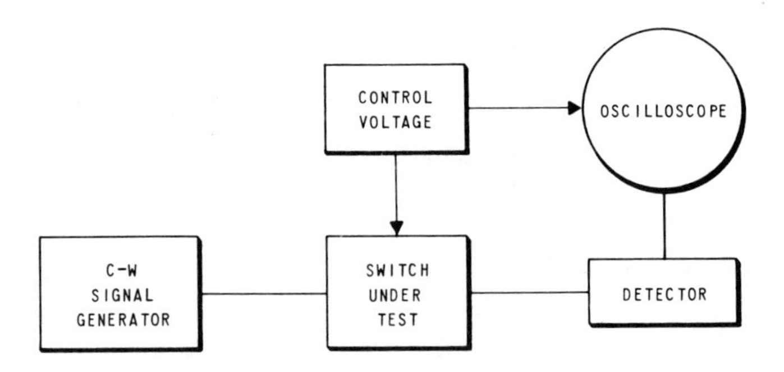

Fig. 11.8. Switching time measurement.

actuates the switch and simultaneously triggers the sweep of the fast oscilloscope. Time is then measured on the calibrated face of the oscilloscope. This circuit will measure *on* time for all switches, and *off* time for all but gas switches.

Gas switches used in high-power circuits are "captured" by the high-power signal even when the switch is not self-triggered; that is, the high power maintains the ionized state and the switch cannot be shut off, until the high-power pulse is terminated. Even after the pulse is over, a finite time is required for the ionization to "die down" to a level which will not attenuate microwave signals. By definition, the *recovery time* is the time between the end of the high-power pulse and the moment the gas switch attenuation drops to three decibels.

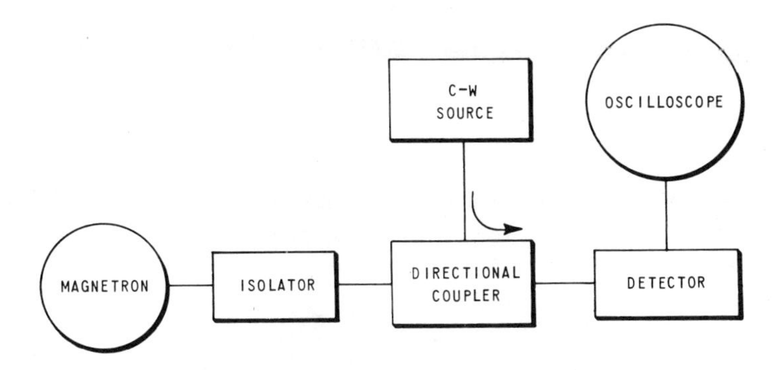

Fig. 11.9. Recovery time measurement.

Measurement of recovery time is shown in Fig. 11.9. The signal from the c-w signal generator is first observed on the oscilloscope. When the magnetron is fired, its repetition frequency is controlled by the sweep circuit of the oscilloscope. During the pulse, there will first be a small leakage blip on the oscilloscope and then (after the tube fires), there will be no signal. After the pulse, ionization starts to decrease and gradually the c-w signal increases. The point where the c-w signal is three decibels below its original value is noted; the time elapsed since the original blip is measured on the calibrated time base of the scope. The pulse width is subtracted from this measurement to give the recovery time of the gas switch.

11.6. POWER HANDLING CAPABILITY

Gas switches are used in high-power applications where breakdown in the gas is a necessary part of the operation. High power is not limited here. Diode switches will not handle high powers since a voltage across the diode in excess of ten volts may cause it to burn out. However, by careful design, it is possible to construct diode switches and insure that the breakdown voltage will not be exceeded, even when as much as 50 kilowatts of peak power exists. Since the diodes themselves absorb practically no power, there is little danger from heating. Ferrite switches, like all ferrite components, have both peak and average-power limitations. There is always some insertion loss in the ferrite material which means some power is absorbed. This causes heating. If the temperature of the ferrite reaches a critical value, called the *Curie temperature,* it ceases to operate. As a practical matter, degradation of operation occurs at much lower temperatures. High peak powers cause erratic behavior in ferrite components. This is called nonlinear operation and is a property of the material and the geometric configuration.

11.7. CONTROL POWER

The amount of d-c power necessary to operate a switch is frequently a deciding factor in the choice of device. Diode switches draw very little current and require low voltages to switch. Typical values may be three to six volts and about 50 milliamperes. This is much less than a watt. Gas switches which are controlled by a d-c discharge require approximately ten to 20 watts of average power with much higher peak powers. Ferrite switches and gas switches which are magnetically controlled require one to ten watts of power.

11.8. DUPLEXERS

In many microwave systems, it is desirable to use one antenna for both transmitting and receiving. The antenna is connected to the transmitter while it is putting out a signal and is then connected to the receiver between pulses. This is called *duplexing,* and the switch or switching circuit which accomplishes it is a *duplexer.* Duplexing is not restricted to pulsed signals, but may also be accomplished in c-w systems. In this case, the transmitter is always connected to the antenna, but the antenna must send its received signals to the receiver.

An elementary duplexer for both pulse and c-w is shown in Fig. 11.10. The transmitter and receiver are connected to opposite arms of a magic tee; the antenna and a matched load are connected to the other two arms. Power from the transmitter divides between the load and the antenna, but is isolated from the receiver arm by the magic tee properties. Similarly, during reception, half the power from the antenna goes to the receiver and the other half is absorbed in the isolator. The disadvantage of this duplexer is a six decibel loss, three on transmission and three on reception. However, the duplexer is simple, lightweight, and requires no additional circuitry or control voltages. As mentioned above, it works for both c-w and pulsed signals.

Gas tubes are widely used as duplexers in pulsed systems because they are triggered by the high-power signal and consequently do not need special synchronized control systems. A simple circuit called a branched duplexer is shown in Fig. 11.11. It has two gas switches which are known as *TR* and *ATR* tubes. (TR stands for "transmit-receive"

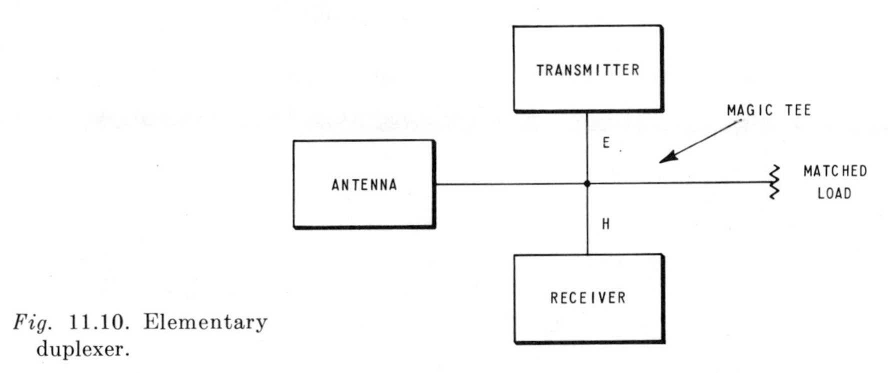

Fig. 11.10. Elementary duplexer.

and ATR is "ante-transmit-receive.") When a pulse comes from the magnetron, it fires both tubes. The ionized gas presents a short circuit at the wall of the waveguide so that, to the pulsed signal, there is no apparent discontinuity. All power goes to the antenna, except for some small dissipation due to arc loss. On reception, the tubes are not fired. The ATR is a quarter of a guide-wavelength long and is terminated in a

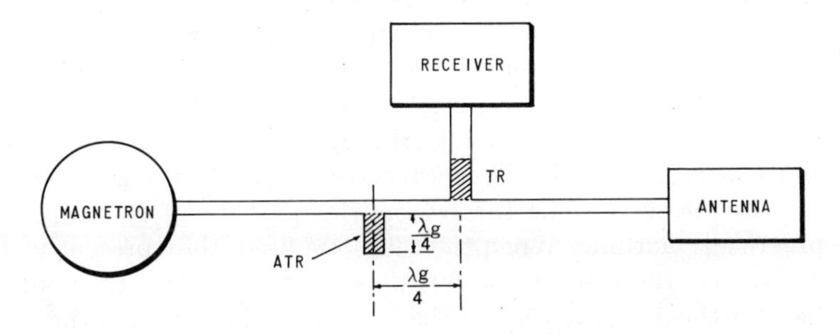

Fig. 11.11. Branched duplexer.

short circuit. This appears as an infinite impedance at the waveguide. To the received signal, this infinite impedance (plus the impedance looking toward the magnetron which cannot make the infinite impedance "more infinite") appears as a short circuit a quarter of a guide-wavelength nearer the antenna where the TR is located. Thus, to the received signal, the magnetron path is shorted out and all power goes to the receiver.

The circuit is limited to a narrow band of frequencies because of the need for sections a quarter of a guide-wavelength long. Since the only receiver protection from the transmitter comes from the fired TR, the isolation through this switch is a limit on high power. Thus, if the receiver crystal cannot tolerate more than 100 milliwatts peak input, and if the isolation of the fired TR is 60 decibels, the peak output of the magnetron must not exceed 100 kilowatts.

A duplexer arrangement which permits operation over a broader frequency range and at higher powers is shown in Fig. 11.12. This is called

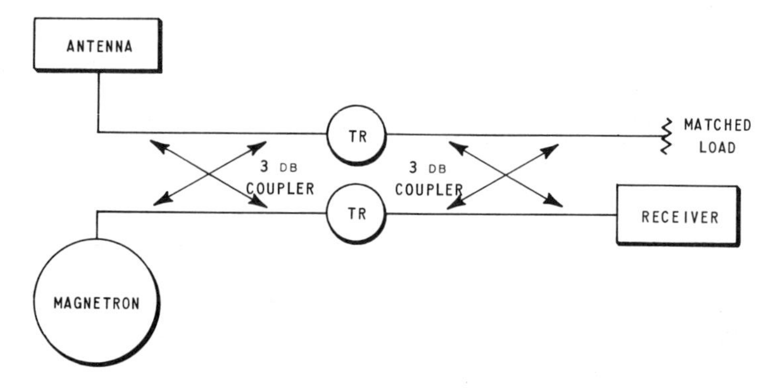

Fig. 11.12. Balanced duplexer.

a balanced duplexer. It uses two short-slot couplers and two TR tubes. The two tubes are usually joined together and have a small hole in their common wall so that the gas pressure is equalized in both. Because of the 90° phase difference in path lengths through the short-slot couplers, a signal into any arm goes diagonally through the circuit, and the other two arms are isolated from it. Thus a received signal at the antenna goes directly to the receiver, while the transmitter port and terminated port receive practically nothing. When the tubes are fired, they present identical short circuits. Because of the 90° phase difference, all the reflected power goes out the arm adjacent to the input arm. Power from the transmitter goes to the antenna. Any leakage through the fired tubes goes diagonally through the circuit just as when the tubes are unfired. Thus, leakage from the magnetron goes to the matched termination so that there is additional protection for the receiver.

It is obviously important that the gas switches fire quickly when the transmitted pulse reaches them, both for receiver protection and to get all the energy to the antenna. To insure rapid firing of the TR tube, an electrode is added to it and a high voltage applied. This voltage is insufficient to fire the tube but produces some local ionization so that the tube will fire more rapidly when a pulse above 100 milliwatts reaches it. The electrode is referred to as an *ignitor* or *keep-alive*. The ATR doesn't have a keep-alive, but all TR tubes do.

11.9. FERRITE DUPLEXERS

The nonreciprocal properties of ferrite materials are used to make duplexing circuits which have no switches and will work for both c-w and pulsed signals. The heart of the duplexer is the ferrite circulator, which was mentioned briefly in Sec. 8.10. The simplest form and most

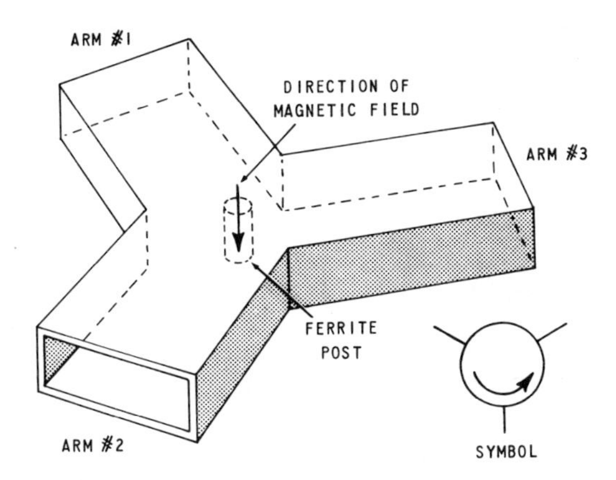

Fig. 11.13. Ferrite circulator.

practical circulator consists of a symmetrical three-port junction that has a ferrite post or disc at the center with a magnetic field applied to the ferrite perpendicular to the plane of the junction. A waveguide version of this circulator is shown in Fig. 11.13. The three arms in the H-plane junction are 120° apart, and the ferrite is at the center of the junction. Suitable matching steps or posts are also placed in the junction so that all three arms are matched when an external magnetic field is applied. Ideally, all power fed into arm No. 1 emerges at arm No. 2, all into arm No. 2 emerges at arm No. 3, and all into arm No. 3 emerges at arm No. 1. In practice, there is a small amount of insertion loss in these directions, varying from 0.1 to 0.5 decibel. The isolation in the reverse direction is not perfect, but varies from 20 to 40 decibels. Coaxial circulators resemble the waveguide type in design and performance.

The ferrite circulator by itself is a simple duplexer. A transmitter, antenna, and receiver are each connected to one arm of the circulator in the proper order. Then the transmitted signal goes to the antenna, while a received signal at the antenna goes to the receiver. (This is illustrated in Fig. 11.14a.) However, if the receiver is not perfectly matched, reflections from the receiver will then go back to the transmitter. To isolate the transmitter from reflections, a four-port circulator should be used. This can be formed from two three-ports, as shown in Fig. 11.14b. A

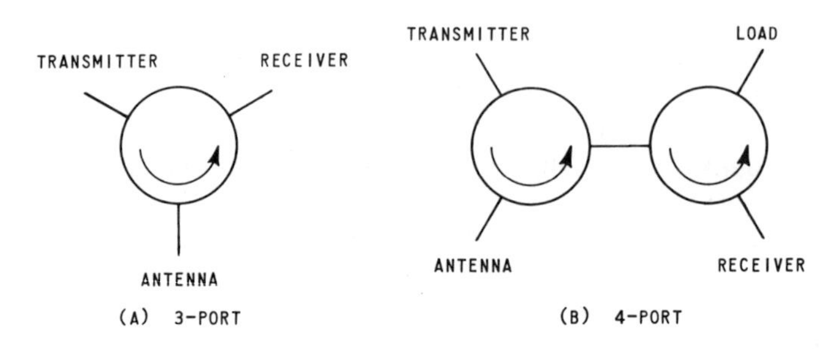

Fig. 11.14. Ferrite duplexer.

matched load is placed on the fourth arm, between the receiver and transmitter.

11.10. CRYSTAL PROTECTION

During operation of a microwave system that uses a gas duplexer, the TR tubes prevent reflections off the antenna from getting to the receiver crystal and causing burn-out. However, when using a ferrite duplexer, any reflection of pulse power from a less than perfect antenna goes directly to the receiver. In order to prevent burn-out, some sort of crystal protector, such as a switch synchronized with the pulse, is necessary. Since the reflection is much less than the transmitter signal, it doesn't require a high-power switch. For this application, diode switches have been found satisfactory.

Another type of switching circuit used to prevent burn-out takes advantage of the nonlinear effect in ferrites mentioned at the end of Sec. 11.6. By proper choice of material and geometry, it is possible to fabricate a component which passes low-power signals but absorbs high-power signals. This is called a *ferrite limiter*. A suitable limiter placed between the receiver and the duplexer will reduce high-power reflections to acceptable values.

When a system is stationed in a busy radar environment, there is always danger that a transmitter in the neighborhood will send a signal directly at it. Because of the short distance between the two radars, the received signal will burn out the crystal unless special precautions are taken. Unfortunately, when the system is in operation, the receiver must be ready to accept signals, and the danger of strong signals must be risked. However, when the system is shut down, the receiver crystal is still susceptible to strong signals. To protect the crystal during periods of shut-down, a simple switch of the type illustrated in Fig. 11.1 is used. Needless to say, the crystal protector must be switched back to the *on* position when the system is put back into operation.

11.11. DIPLEXERS

In Sec. 9.15, diplexers were mentioned briefly as one of the filter applications. A diplexer is a circuit which separates signals by frequency.

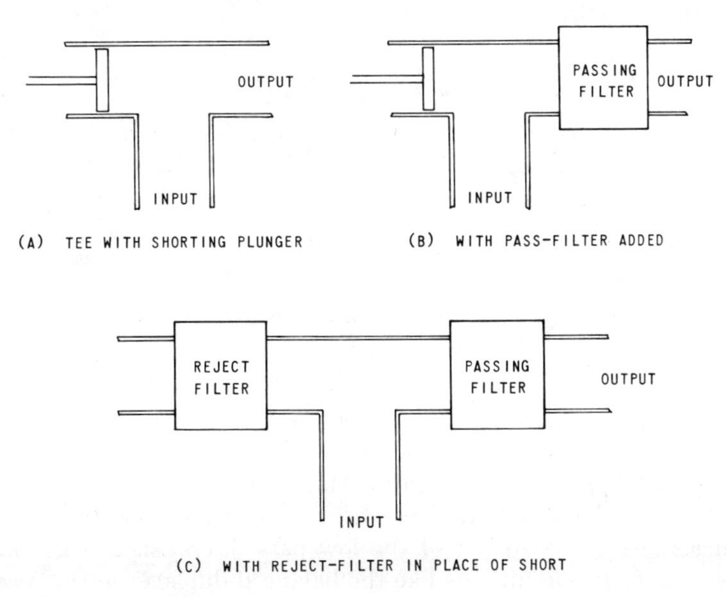

Fig. 11.15. Development of a diplexer.

Effectively it is a three-port network. Two different frequency bands are fed to the input, and each frequency band emerges from a different output arm. Filters are a necessary part of a diplexing circuit.

In order to understand the operation of a simple diplexer, it is necessary to consider a tee with a short circuit in one side arm. Figure 11.15a shows such an E-plane tee with a short-circuiting plunger in one of the arms. At any specific frequency, it is possible to find a position of the plunger so that the remaining two ports are connected with little reflection. With some matching, it is possible to have a low VSWR over a band of frequencies.

If in Fig. 11.15a, a filter which passes the frequency band of interest is put anywhere in arm No. 2, it will not affect the operation. This is shown in Fig. 11.15b. Now, a filter which rejects the frequencies looks like a short circuit. If this type of filter is substituted for the plunger with its short circuit at the point where the plunger was, there is still no change in operation. This is shown in Fig. 11.15c. Looking at Figure 11.15c, it is obvious that there can exist a different frequency band which the filter in arm No. 2 will reject and the one in arm No. 3 will pass. For this band, a signal in arm No. 1 will come out arm No. 3, while for the

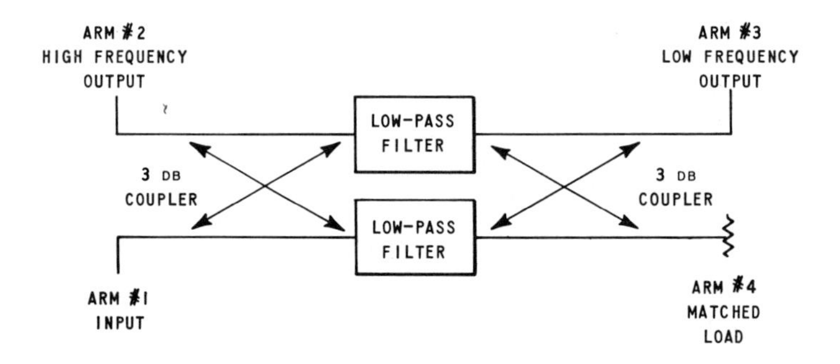

Fig. 11.16. Balanced diplexer.

original band, a signal in arm No. 1 will come out arm No. 2. This is a diplexer.

A *balanced diplexer* is shown in Fig. 11.16. The similarity between this circuit and the balanced duplexer of Fig. 11.12 should be noted. At frequencies below the cut-off of the low pass filters, signals go through the filters, and the circuit acts like the balanced duplexer in the receiving state. Thus, if these frequencies are fed into arm No. 1, they emerge at arm No. 3. At frequencies above the cut-off frequency, the filters look like short circuits. Thus, the circuit behaves as the balanced duplexer when the TR tubes are fired. When these higher frequencies are fed into arm No. 1, they emerge at arm No. 2.

11.12. MODULATORS

Any switch can be used as a modulator. The rate of modulation is determined by the type of switch; mechanical switches can easily be operated at 60 cycles and may be used perhaps up to 1000 cycles. Ferrite switches which require a change of magnetic field can be switched comfortably at kilocycle rates and, in some circuits, up to a megacycle. Diode switches which require only a voltage change have been switched at speeds of a few nanoseconds; that is, at frequencies in excess of a hundred megacycles.

With conventional techniques, a modulator adds power to the carrier signal. When one of the switches described above is used as a modulator, however, it simply turns the power on and off at a prescribed rate and manner. Thus, the modulated microwave signal has less power than the unmodulated signal.

QUESTIONS AND PROBLEMS

11.1. How can complete reflection be obtained from a post being used as a switch in a waveguide?

11.2. Explain the term "isolation" as it is used in connection with multiple position switches. Show with a block diagram how this property of a switch might be measured.

11.3. Show with a block diagram how two switching diodes and a square-wave generator can be used to send modulated waves up each of two coaxial lines from a single c-w signal generator. Use a minimum of additional parts.

11.4. Define the term "arc loss." Explain how it can be measured.

11.5. Explain in detail the operation illustrated in Sec. 11.8. Draw a diagram of what one would expect to see on the scope.

11.6. Explain how it is possible for small diodes to switch peak power levels of up to 50 kw.

11.7. Distinguish between duplexers and diplexers.

11.8. What limits the useful bandwidth of the TR tube? Why is a keep-alive voltage sometimes applied to it?

11.9. Explain the operation of and the reason for the use of balanced duplexers.

11.10.

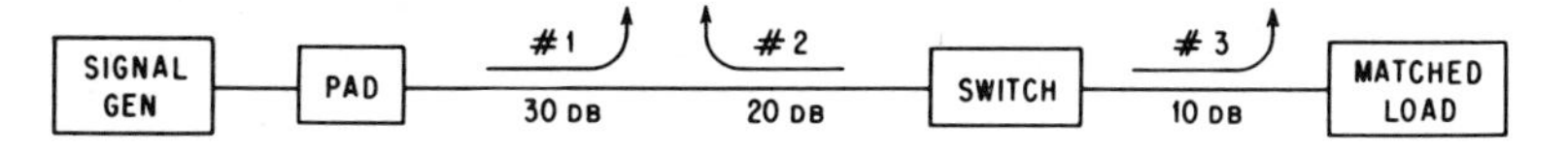

Fig. P 11.10

Data:

With switch closed			switch open
Coupler #1	(30 db coupler)	reads 0 dbm	0 dbm
Coupler #2	(20 db coupler)	reads 9.6 dbm	−4.1 dbm
Coupler #3	(10 db coupler)	reads 0.2 dbm	19.6 dbm

a. Find the following with switch closed:
   1. db isolation of the signal generator *ans.* 10 db
   2. Isolation of switch *ans.* 19.8 db
   3. Power dissipation in switch *ans.* 77.5 mw
   4. Power into load *ans.* 10 mw

b. Find the following with switch open:
   1. Power into the load *ans.* 912 mw
   2. VSWR of the switch *ans.* 1.5
   3. Insertion loss of the switch. *ans.* .4 db

(Assume in all of the above that the power absorbed by the couplers is negligible. Is this really justified?)

12

ANTENNAS

At microwaves, as at lower frequencies, the power emitted from the transmitter must be radiated into free space in the form of electromagnetic waves. On the receiving end, electromagnetic waves must be intercepted and fed into the transmission line running to the receiver. The component which radiates and intercepts, is, of course, the antenna. As was mentioned in the preceding chapter, the same antenna may serve both purposes, when the transmitter and receiver are at the same site.

The antenna can be thought of as a matching network which couples the transmission line to free space with minimum reflection and loss. In addition, the antenna can be shaped to propagate the electromagnetic wave in a particular direction and to present, depending upon the application, a broad or narrow beam.

12.1. RECIPROCITY

Antennas are reciprocal. An antenna used with a transmitter has the same characteristics and performance as it would have when used with a receiver. Thus, the VSWR looking into the terminals of an antenna is the same whether it is used as a transmitting antenna or as a receiving antenna. The gain, beamwidth, side-lobe level, and other characteristics are the same for both applications.

This reciprocity is convenient in that measurements may be made using the antenna while it is either transmitting or receiving, whichever is simpler. Since transmission is usually at high power levels, and reception at very low power levels, it is apparent that power level also makes no difference on performance and characteristics (assuming no breakdown). Measurements are made at the most convenient power level and at transmission or reception, based on convenience.

12.2. GAIN

An antenna which radiates in all directions equally is called an *isotropic* radiator or source. There is no such antenna, because every antenna exhibits some directive properties. However, the isotropic antenna is a convenient reference point, and thus the gain or directivity of a real antenna is expressed as the increase in power radiated in a given direction compared to the power radiated by the fictitious isotropic antenna, assuming the same total power in both cases. Obviously if an antenna has directivity, the gain is a function of direction from the antenna. Thus it is referred to as the *gain function* when discussing variation in gain with direction. The maximum value of the gain function is referred to simply as the *gain* and is designated G. It is measured in decibels. A matched transmitting antenna with a gain of 12 decibels would put out a signal (in the direction in which the signal is maximum) 12 decibels greater than a signal from an isotropic source which is fed by the same transmitter.

When an antenna is used for receiving, and a plane wave impinges on it, the amount of power intercepted is proportional to its area. The cross section then is proportional to the gain:

$$A = \frac{\lambda^2 G}{4\pi} \tag{12.1}$$

where A is the area of cross section, G is the gain, and λ is the free-space wavelength. Equation (12.1) applies to a matched antenna whether it is used for transmitting or receiving.

From Eq. (12.1), it is apparent that for a fixed size of antenna, the gain is inversely proportional to the square of the wavelength. The antenna dimensions then should be expressed in wavelengths when comparing antennas at two different frequencies. A round antenna which is 30 wavelengths in diameter at 1000 megacycles should have the same gain as a 30-wavelength antenna at 10,000 megacycles. This is evident by solving Eq. (12.1) for G.

$$G = \frac{4\pi A}{\lambda^2} \tag{12.2}$$

Since A is proportional to the square of the diameter, and the diameter is 30 wavelengths, G is the same for both frequencies.

12.3. POLARIZATION

Electromagnetic waves consist of an electric or voltage field and a magnetic field. These are always perpendicular to each other. By definition, the *polarization* of an electromagnetic wave is the direction of the electric field. If this field is perpendicular to the earth, the wave is *vertically* polarized; if the voltage field is horizontal, the polarization is *horizontal.* A field may be at some intermediate angle between zero and 90 degrees, in which case the field is usually said to have both horizontal and vertical components, although it would be just as correct to say that the field is at a specified angle from the vertical or horizontal. Figure 12.1 illustrates a voltage vector which is polarized an angle θ from the vertical. From the diagram, it is obvious that the vertical component is $E \cos \theta$, and the horizontal component is $E \sin \theta$, where E is the amplitude of the voltage vector of the original wave. The power in the horizontal polarization is the square of the voltage or

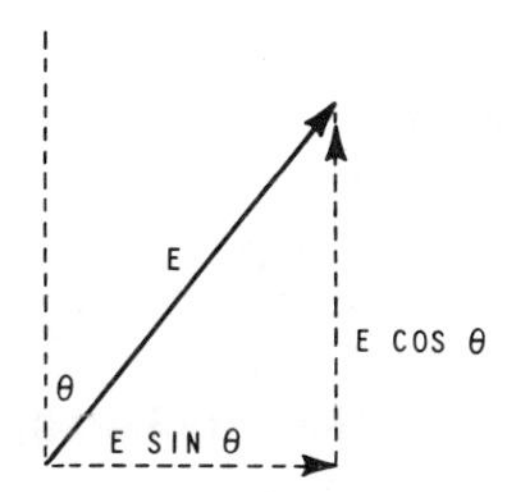

Fig. 12.1. Components of polarization.

$$P_H = E^2 \sin^2 \theta \tag{12.3}$$

Similarly, the power in the vertical polarization is

$$P_V = E^2 \cos^2 \theta \tag{12.4}$$

The sum of these two is

$$P_H + P_V = E^2(\cos^2 \theta + \sin^2 \theta) = E^2 \tag{12.5}$$

which is, of course, the power in the original wave. Thus, a wave polarized at an acute angle to the vertical may be considered equivalent to two waves travelling together, one horizontally and the other vertically polarized.

At the beginning of this chapter, it was pointed out that an antenna can be thought of as a matching transformer between the transmission line and free space, and that the antenna also directs the radiation. To this can be added that the antenna determines the polarization of the radiated field. If an antenna launches a horizontally polarized wave, the antenna is said to be *horizontally polarized.*

A horizontally polarized antenna launches a wave which has no vertical components. By reciprocity then, such an antenna will receive no vertical components. Thus, for maximum efficiency, the transmitting and receiving antennas should have the same polarization. If the receiving antenna is polarized at 90° with respect to the transmitting antenna, it will receive no signal. The antennas are said to be *cross polarized.* Maximum reception occurs when the antennas have the same polarization, minimum when they are cross polarized. In between, the amplitude varies as the cosine of the angle between the two antennas.

Sometimes it is necessary to transmit signals to a point where the polarization of the receiving antenna is unknown and may be at any angle. A common solution is to transmit *circularly polarized* waves. In circular polarization, the polarization of the wave rotates at the r-f rate. At any moment in time, the voltage vector has a component correctly oriented for the receiving antenna and one perpendicular to this. Thus, half the power will be received, regardless of the polarization of the receiving antenna. Similarly, when searching for signals of random polarization, a circularly polarized antenna can be used on reception.

An antenna which has a single direction of polarization (as opposed to circular) is said to be *plane polarized.* Such an antenna can be used to determine the polarization of a transmitted signal simply by rotating it until the received signal is maximum. If the amplitude of the received signal stays constant as the plane-polarized receiving antenna is rotated, then the transmitted signal must be circularly polarized. If the amplitude decreases to zero in one polarization and is maximum 90° from this, then the direction of polarization is that at which the amplitude is maximum. Sometimes the signal has no zero polarization and is not constant, either. This is elliptical polarization, which, in general, is not deliberately transmitted, but is imperfect circular polarization.

12.4. BEAM PATTERN

If a waveguide is left open, a wave propagating in the guide will radiate from the open end. Since the waveguide impedance differs from that of free space, not all the energy will be radiated, but some will be reflected back toward the generator. The phase and amplitude of the mismatch can be determined in the usual way. A simple matching iris or post can then be used to match the open waveguide so that all the power is in fact radiated. The open guide is thus an antenna, but from Eq. (12.2), it is evidently a low gain antenna. Thus, if the waveguide is 0.4 by 0.9 inches (X-band), A is 0.36 square inch. At 10,000 megacycles, λ is 1.18 inches. From Eq. (12.2),

$$G = \frac{4\pi(0.36)}{1.39} = 3.25 \approx 5 \text{ db}$$

This assumes a uniform distribution across the aperture so that the effective area is the actual area. As a matter of fact, it is obvious that the aperture distribution is not uniform. The voltage is maximum at the center of the waveguide and is zero at the side walls. The same distribution is carried to the aperture. The result is to produce an effective area which is less than the 0.36 square inch indicated. Consequently, the gain also will be less than the five decibels mentioned above. The concept of effective area will be treated later in this chapter.

Equation (12.2) indicates that the larger the area, the larger is the gain. Since a larger gain means that more energy is transmitted in a preferred direction, it must of necessity imply that this extra energy is taken from other directions so that the beam of energy becomes narrower. In general, the larger the aperture becomes in terms of wavelengths, the narrower is the beam.

The relative intensity of the radiation as a function of direction can be plotted in polar coordinates, as shown typically in Fig. 12.2. This is called a *beam pattern,* or simply, a pattern of the antenna. The beam pattern for the E-plane is usually different from that in the H-plane. This can be easily understood by considering an antenna consisting of a single vertical wire, half a wavelength long. Obviously, it will radiate equally well in all horizontal directions so that its azimuth pattern is a circle. In elevation, however, it may have zero radiation off the ends and maximum radiation straight ahead. Since the wire is vertical, the voltage field is vertical and the magnetic field is horizontal. Thus, the pattern in the horizontal plane is in the plane of the magnetic field and is called the *H-plane pattern.* Similarly, the elevation pattern is the *E-plane pattern.* Of course, if the wire were horizontal, the azimuth pattern would be the E-plane pattern.

The pattern of Fig. 12.2 indicates that there is a favored direction of propagation for the antenna in the indicated plane of polarization (in this case the E-plane). The other plane of polarization may be similar or may be entirely different. The largest lobe in the figure is called the *main lobe;* smaller lobes are called *sidelobes.* If there is a lobe 180° from the main lobe, it is called a *back lobe.*

In many types of microwave antenna, a small radiator illuminates a reflector or lens which then redirects the energy into free space. The small primary radiator has a beam pattern of its own which is called the *primary pattern* of the composite antenna. The pattern of the whole antenna is strictly the *secondary pattern,* but is usually called simply the *pattern.*

The width of the main lobe of the pattern at the half-power points is called the *beamwidth* of the antenna. Just as the E- and H-planes have different patterns, so also do they have different beamwidths. It should be noted that beam patterns can be plotted in decibels against angular direction (as in Fig. 12.2) in power, or in voltage. The half-power points are the three decibel points on the decibel plot, the 0.5 points on the power plot, or the 0.707 points on the voltage plot. The chart is always normal-

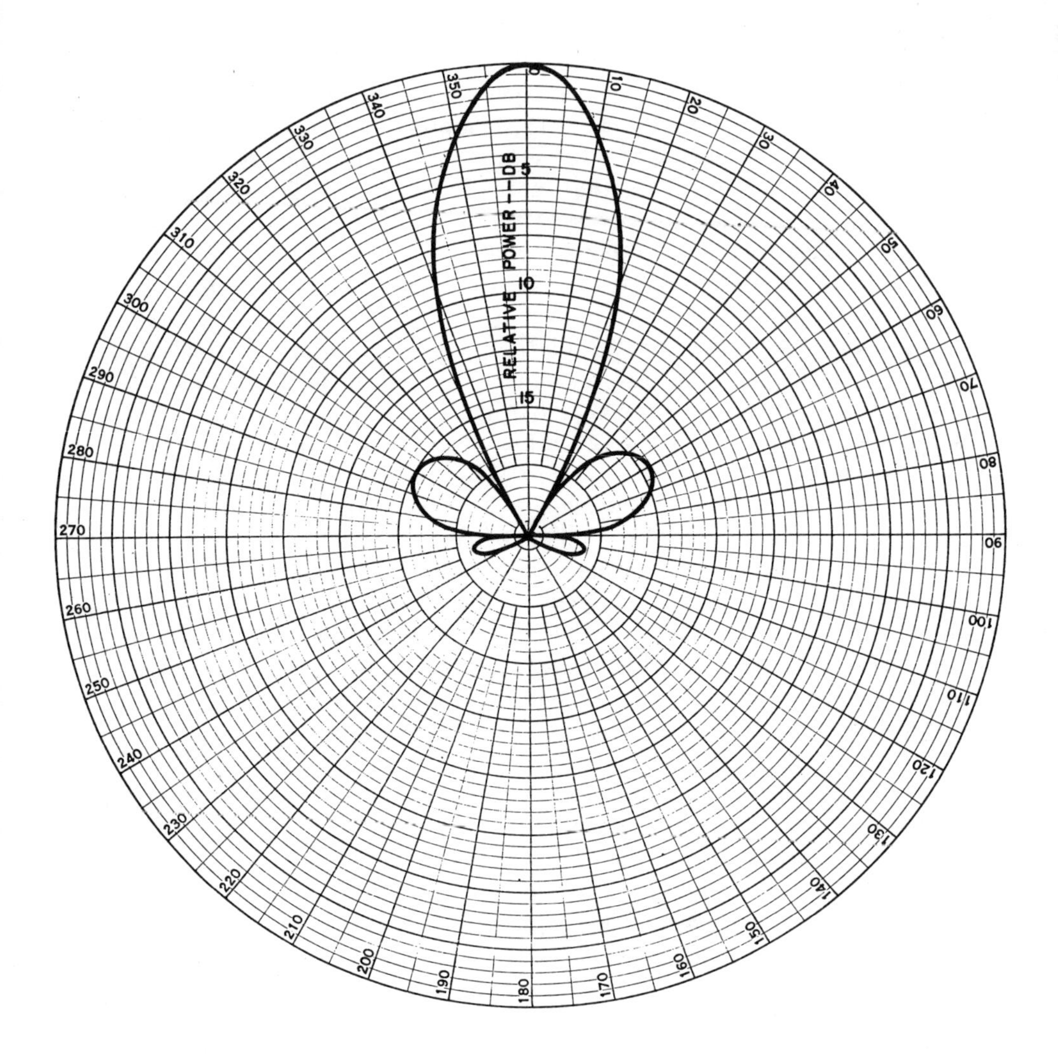

Fig. 12.2. Beam pattern.

ized so that the maximum point is unity or zero decibels. In Fig. 12.2, the beamwidth is 20°.

12.5. SIDELOBES

In order to achieve maximum gain from an antenna, the aperture must be illuminated uniformly and in phase by the feed to the antenna.

When this happens, the gain of the antenna is a function of the area of the aperture and is given by Eq. (12.2). There is a disadvantage in this arrangement, however, in that the discontinuity presented by the edge of the antenna occurs at a point of high intensity, which produces sidelobes. Sidelobes, of course, are undesirable since they may give erroneous indications of direction. If the discontinuity represented by the antenna edge is at a point of lower intensity, the sidelobes would be reduced. Thus, typically it is customary to strive for a tapered illumination of the aperture, with maximum intensity at the center, tapering to ten decibels down at the edge. This reduces the sidelobes but also reduces the gain, since the tapered radiation from the aperture is equivalent to uniform radiation from a smaller aperture.

For a perfectly symmetrical antenna, the pattern should also be symmetrical. However, mechanical tolerances would be too stringent to accomplish this. In practice, there is always some asymmetry which shows up particularly in unsymmetrical sidelobes.

For a parabolic reflector, uniformly illuminated, the gain is as given by Eq. (12.2), and the sidelobes are down 13 decibels; that is, the peak of the nearest sidelobes is 13 decibels below the peak of the main beam. If the parabola is illuminated so that the circumference is ten decibels below the center, the sidelobes are reduced to 17 decibels below the main beam. The gain is also reduced to about 55 per cent of the value given in Eq. (12.2). The antenna is then said to have an efficiency of 55 per cent.

12.6. BEAMWIDTH

In Sec. 12.4, beamwidth was defined as the width of the beam at the half-power points. Inasmuch as the gain and sidelobe level of an antenna pattern vary with the aperture illumination, it might be expected that the beamwidth would also vary, and this is indeed the case. In general, the beamwidth is inversely proportional to the aperture width. Also, if the aperture is uniformly illuminated, it is effectively larger than the same aperture with tapered illumination and thus will have a narrower beam.

A parabolic dish antenna with tapered illumination down 10 db at the edges has a beamwidth given by

$$\mathrm{BW} = \frac{70\lambda}{D} \tag{12.6}$$

where λ is the free-space wavelength, and D is the diameter of the dish. The beamwidth is the same for both E-plane and H-plane patterns, since the same illumination is assumed for both. In practice, it would be difficult to build a feed with a primary pattern which illuminates the dish so that the taper is the same in all directions from the center out to the perimeter of the dish. Thus, the E- and H-plane patterns usually are not the same.

The simple waveguide antenna described in Sec. 12.4 can be matched by flaring the walls of the waveguide gradually until the impedance of the enlarged guide approximates that of free space. Such a horn antenna has gain which is directly related to the area of the aperture, but since this aperture is not uniformly illuminated, the efficiency is never 100 per cent. In practice, the efficiency of a horn antenna is between 50 and 80 per cent. The aperture is, of course, a large waveguide in which the *voltage* is maximum at the center and zero at the side-walls. In the H-plane then, the *distribution* is maximum at the center tapering to zero at the walls. The H-plane beamwidth is given by Eq. (12.6), just as for a parabola with tapered illumination, but D here is the dimension of the aperture in the H-plane. Since there is no variation in illumination in the E-plane direction, for $TE_{1,0}$ mode excitation, the E-plane beamwidth will be narrower for a given dimension than the H-plane pattern. The E-plane beamwidth is

$$\text{BW} = \frac{51\lambda}{D} \tag{12.7}$$

where D now is the dimension across the horn in the E-plane. Equation (12.7) is also the beamwidth of a uniformly illuminated parabolic dish.

12.7. ANTENNA TYPES

Besides the horn and parabolic dish antennas mentioned thus far, microwave antennas come in a variety of shapes and sizes. Miniature variations of low-frequency antennas are used at microwaves in arrays or as primary feeds to illuminate reflectors. These antennas include dipoles, rhombics, Yagis, and other low-frequency types. Reflectors are also illuminated by configurations which are peculiar to microwaves and have no low frequency counterparts. Typical is the small horn or open waveguide which is used as a feed in many radar installations.

Another microwave antenna is the waveguide slot radiator. When a slot is in the center of the broad wall of a waveguide, as in a slotted line, there will be no radiation from the guide. However, if the slot is located off center, where it will cut lines of current flow, there will be radiation through it. A single slot can be used as the primary feed to illuminate a reflector, or an array of slots may be used as a complete antenna.

The paraboloid or parabolic dish is usually used as a reflector in microwave systems because of two special geometric properties. These are illustrated in Fig. 12.3 in a simple parabola. First, all rays from a fixed point, called the focal point, to the parabola are reflected as parallel rays. The focal point or focus is designated F in Fig. 12.3. Secondly, the sum of the distances from F to the parabola, and then along the reflected ray to some reference plane, PP′, is a constant. Thus, the reflected wave is made up of parallel rays which are all in phase.

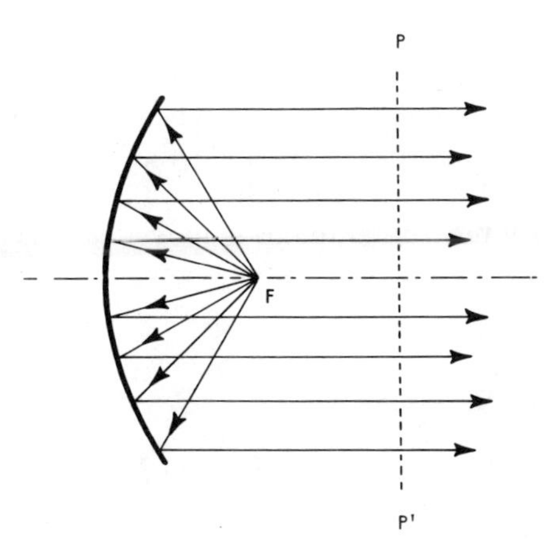

Fig. 12.3. Parabola.

In general, the antenna beam can be shaped to any desired configuration, depending on the application. The larger the dish, the narrower will be the beam. A narrow pencil beam can be used to pinpoint the direction of a target, but it would be difficult to use in locating targets whose positions are random. A pencil beam could scan in azimuth at a fixed elevation, but would miss targets above or below this elevation. A *fan beam* which has a narrow width in azimuth and a wide beam in elevation would be better. Then when a target is found in azimuth, its elevation can be determined more precisely later. Such a fan beam results from an aperture which has a large dimension in the azimuth direction and a narrow dimension in the elevation direction. This could be a horn with the proper dimensions or a strip of a parabolic dish.

12.8. CONICAL SCANNING

When a radar is used to direct guns toward a target, the antenna must indicate the direction of the target quite accurately. In general, a simple pencil beam cannot indicate direction with sufficient accuracy, since the difference in reflected signal between on-target and a near-miss is not usually great enough to detect. Furthermore, when a target is detected, the operator must swing the beam back and forth through it in order to determine when the target is at the peak of the beam. In fact, when the target is first detected, it is impossible to tell whether it is on the peak or off to the side; and if off to the side, it is impossible to tell whether the movement should be to the left or to the right.

One method of solving this problem is called *conical scanning*. The beam is made to nutate about an axis as it scans in azimuth or elevation. Thus, in a plane through the axis the beam will occupy extreme positions indicated in Fig. 12.4. At one point in its nutation, it will occupy position

A, and half a revolution later, position B. If a target is in the direction X in the figure, the signal received will have an amplitude a when the beam is in position A, and an amplitude b, when the beam is at B. Thus, the received signal will be modulated at the rate of nutation. If the target is "dead-ahead," on the axis of nutation, there will be no modulation, since the amplitude remains the same with nutation. The point where the two extreme positions of the beam cross is called the *cross-over point*. It should be from one to three decibels below the peak of the beam.

In operation, when a target is detected somewhere near the beam center, the antenna is moved so that the modulation is decreased. When the modulation reaches zero, the axis of nutation is pointing right at the target.

12.9. MONOPULSE

If instead of a single nutating antenna, two antennas are used pointing in slightly different directions, the lobes would have a configuration similar to that of Fig. 12.4 for conical scanning. Instead of a modulation, there would not be two received signals. On boresight, the two signals would have the same amplitudes, but if the target were off center, the amplitudes would differ.

Again, the two antennas could be pointed in the same direction, that is with beams parallel. If the antennas are separated slightly, a signal from a target would arrive at the two antennas at different times, that is, in different phases, unless the target was right on the line midway between the two antennas.

In both cases, the target direction can be determined accurately without the mechanical motion and its associated problems in conical scanning. By proper circuitry the two signals can be compared automatically in phase or amplitude so that the target will be tracked. Suitable arrangements of two or four antennas can be used to obtain information in both azimuth and elevation.

The systems which use two or more antennas in this way are called *monopulse* or *simultaneous lobing* systems. Their main advantage over conical scanning is they have no moving parts. However, the circuitry is more intricate, and the system is more susceptible to spurious responses.

12.10. SEQUENTIAL LOBING

In conical scanning, the antenna beam nutates continuously around a fixed axis. Figure 12.4 indicated two extreme positions of the beam. Instead of rotating one beam, it should be possible to use two fixed beams which always occupied these same extreme positions. If the beams

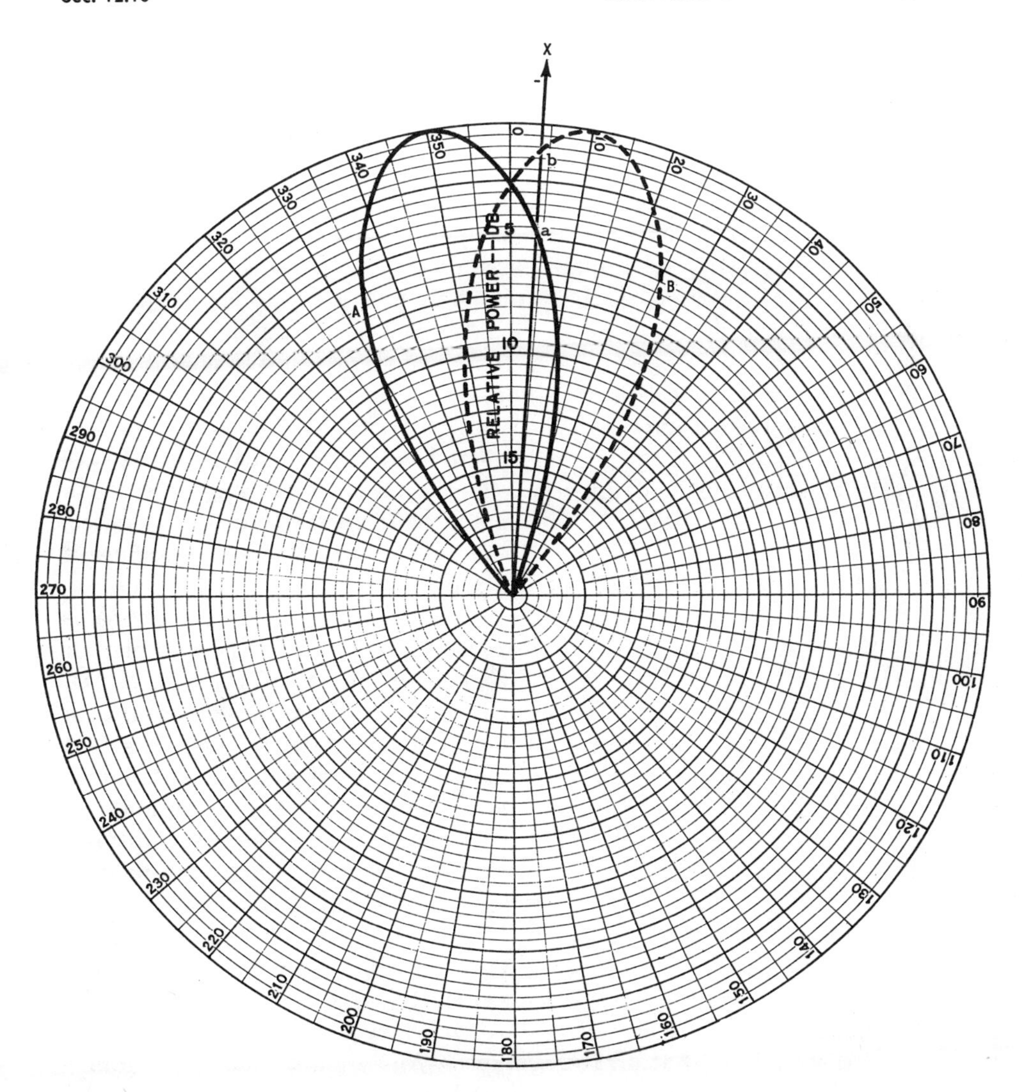

Fig. 12.4. Conical scanning.

are alternately connected to a single receiver, the resultant modulation would indicate direction in the plane of the two beams just as in the case of conical scanning. This method is called *sequential lobing.* In practice, it is common to use four antennas, producing four beams at 90° intervals around the axis of "rotation." The antennas are switched *on* (that is, they are connected to the receiver) sequentially, and the modulation indicates direction of the target. This system has no moving parts, but does require four antennas and suitable switches. Usually diode switches, as described in Sec. 11.2, are used.

12.11. NEAR AND FAR ZONES

The antenna pattern consists of a main lobe and several side lobes with sharply defined minimum points between the various lobes. At a distance from the antenna, assuming it is transmitting, maximum intensity is observed at the peak of the main lobe, and maxima are also observed in the directions of the sidelobes. No signal is detected at the minimum points. However, close to the antenna, this is not the case. There are no sharply defined minima, and the maxima are diffuse. The beam pattern of an antenna, as discussed in Sec. 12.2, refers to the far-zone pattern. Close to the antenna, there is no definite pattern, since rays seem to be traveling in all directions. As they get farther from the antenna many of these rays fade out or cancel each other, and then a true pattern is formed.

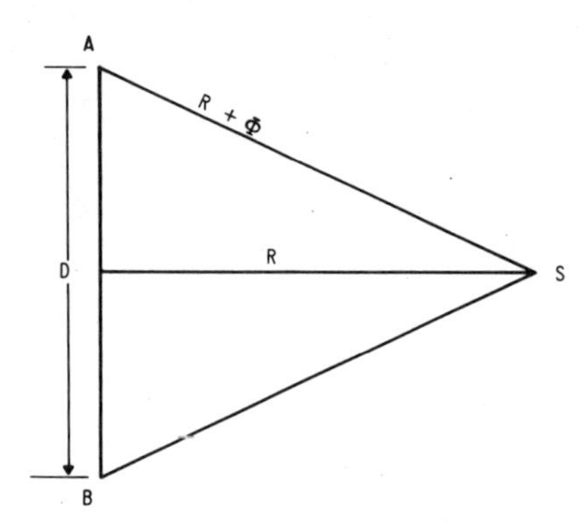

Fig. 12.5. Minimum range determination.

The near zone of the antenna is called the *Fresnel region*, and the far zone is called the *Fraunhofer region.* Obviously, there is no definite line of demarcation between them. Pattern measurements ideally should be made far enough away from the antenna to simulate actual target conditions. However, if an antenna is to be used for ten mile ranges, for example, it is usually difficult to make pattern measurements over the distance. Consequently, measurements are made at a distance which is convenient, but far enough to be a good indication of what would occur at greater distances. An accepted value of the minimum distance for pattern measurements is

$$R = \frac{2D^2}{\lambda} \tag{12.8}$$

where R is the minimum range, D is the diameter or width of the aperture and λ is the free-space wavelength. For example, assume the antenna is a parabolic dish ten feet in diameter to operate at a wavelength of one inch. From Eq. (12.8), R is 28,800 inches or 2400 feet. This is a minimum distance. In practice, pattern measurements are made on this type of antenna at distances of about one mile. Again, assume the antenna is a horn with its widest dimension five inches, to be used at the same frequency. Here R is 50 inches, and, at this length, measurements can be made indoors quite comfortably.

Equation (12.8) can be appreciated by considering Fig. 12.5. In this figure, S represents a source, and the line AB represents an antenna with an aperture D. The distance from the source S to the center of the antenna AB is R. The distance to the edge of the antenna is the line AS which is longer than R by some phase difference ϕ. By the Pythagorean theorem,

$$\left(\frac{D}{2}\right)^2 + R^2 = (R + \phi)^2 = R^2 + 2R\phi + \phi^2 \tag{12.9}$$

Ideally, the phase at the center of the antenna should be the same as at the edge, but this is obviously impossible. For practical systems, the phase difference should not exceed $\lambda/16$. If in Eq. (12.9), $\phi = \lambda/16$, ϕ^2 is negligible and can be omitted. This then becomes

$$\frac{D^2}{4} + R^2 = R^2 + \frac{R\lambda}{8} \tag{12.10}$$

Solving for R results in Eq. (12.8). If ϕ is smaller than $\lambda/16$, R will be larger, since R varies inversely as ϕ.

12.12. FREQUENCY

Antenna characteristics are usually frequency-dependent, as should be evident from the equations covered in this chapter. Thus, from Eq. (12.2), the gain of an antenna of fixed area increases as the square of the frequency. Equations (12.6) and (12.7) indicate that as the frequency increases, the beam becomes narrower. Finally, Eq. (12.8) indicates that for a fixed antenna, the transition between near and far zones also increases with frequency.

In practice, antennas are designed so that the minimum gain in the frequency range will exceed or equal the gain specification. Similar considerations hold for beamwidth. In making measurements, the antenna is usually far enough from the source to exceed the distance given by Eq. (12.8) at the highest frequency.

12.13. r-f DARK ROOM

When making any kind of antenna measurements indoors, by its very nature the antenna radiates energy which hits walls, furniture, and personnel and then reenters the antenna to give erroneous readings. It is important, therefore, to minimize such reflections and to take precautions to prevent them from affecting the readings. Indoor antenna measurements are usually made in a room lined with material which absorbs microwave signals almost completely, thus reducing reflections to a negligible value. Such a room is called a microwave dark room from its analogy to a photographic dark room where all light is absorbed by the walls.

All VSWR measurements on an antenna must be made in a dark room or with the antenna radiating into free space. It is evident that reflections which return to the antenna will be picked up by the slotted-line probe to give erroneous readings. In the dark room, stray energy is absorbed by the walk, ceiling, and floor, so that spurious reflections are avoided.

Similarly, if pattern measurements are made indoors, but not in a dark room, reflections cause erroneous readings. Energy in a sidelobe of the transmitting antenna can be reflected so as to be picked up in the sidelobe of the receiving antenna and appear to be additional energy in the main lobe.

12.14. VSWR MEASUREMENTS

Measurements of VSWR on an antenna are made in the same manner as on any other microwave component, with the added precaution of preventing external reflections. Ordinarily, an antenna is designed and engineered to meet required specifications of gain, pattern shape, and sidelobe level. To achieve this, changes are frequently made on the shape of the primary antenna and the geometry of the composite antenna. Now VSWR is measured, and, if it is not within specifications, it is usually corrected within the feed to the primary antenna. Any changes in this feed have no effect on the other antenna parameters, whereas changes outside the feed almost certainly will affect the pattern.

There are exceptions to the rule of correcting VSWR inside the feed. When the focal point of the parabola, where the primary feed horn is located, lies right in the center of the beam, energy reflected from the vertex of the parabola is intercepted by the feed horn with a resultant degradation in VSWR. This can be prevented by directing the energy from the vertex elsewhere and is accomplished by putting a small plate at the vertex. This plate is shaped so that reflections from it do not go back to the feed horn. The plate is called a *vertex plate* or a *spoiler*, because, it spoils the pattern slightly.

12.15. GAIN MEASUREMENTS

The gain of an antenna is usually measured by comparing the antenna to a calibrated "standard" antenna. (Figure 12.6 shows the measuring set-up.) The transmitter is fed to a radiating antenna which is pointed toward the test antenna. The antenna under test is moved in azimuth and elevation until the received signal is peaked to a maximum. The maximum received power P_1 is noted. Then the receiving antenna is replaced by the calibrated standard, usually a horn, and the same detection system is connected to the standard. This antenna is peaked, and the new maximum received power, P_2, is noted. To simplify calculations,

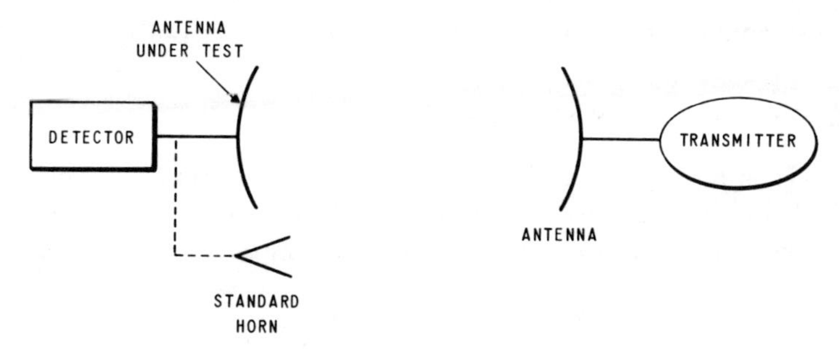

Fig. 12.6. Gain measurement set-up.

P_1 and P_2 are noted as decibel readings on an arbitrary scale. Then the gain of the antenna under test is

$$G = P_1 - P_2 + G_s \tag{12.11}$$

where G is the gain in decibels, G_s is the gain of the standard in decibels, and $P_1 - P_2$ is the difference in decibels between the two received signals. For example, assume that with the test antenna connected, the detector output meter is adjusted to read full scale or zero decibels. Now with the standard horn in place, the output is down seven decibels or reads minus seven decibels. Thus $P_1 - P_2 = 7$ db. If the standard horn has a gain of 15 decibels, the test antenna is seven decibels better and has a gain of 22 decibels.

If the detector is not square law, the gain difference (seven decibels in the above example) may be in error. The error will be small for small differences, but can be appreciable for differences in excess of ten decibels. To avoid this difficulty especially if the detector response is unknown, a calibrated attenuator can be used between the higher gain antenna and the detection system. The attenuator can be adjusted to make the two responses identical. The gain is then

$$G = G_s \pm \alpha \tag{12.12}$$

where α is the setting of the attenuator. The plus sign is chosen when the test antenna has a higher gain than the standard and the attenuator is thus used with the test antenna. The minus sign is valid when the attenuator is used with the standard.

It is important that both the test antenna and the standard are well matched; otherwise, loss due to reflection will appear as a reduction in gain. The detector should also be matched, and the same detection system used for both antennas.

The distance between the transmitting site and the receiving site must be far enough so that slight changes in position of the two antennas will not cause changes in readings. The criterion of Eq. (12.8) is satisfactory.

12.16. PRIMARY GAIN STANDARD

It is possible to measure the gain of an antenna directly, if two identical antennas are available. This method is usually selected to calibrate horns which are to be used as primary standards, but it can also be

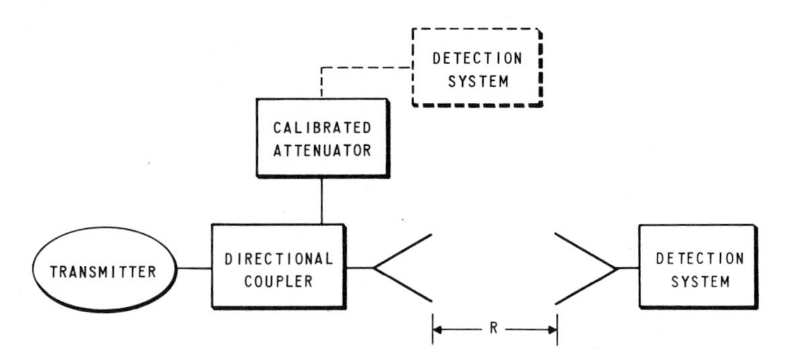

Fig. 12.7. Primary gain standard measurement.

used for any kind of antenna. (A set-up is shown in Fig. 12.7.) The two antennas are separated by a distance at least as great as $2D^2/\lambda$, as indicated in Eq. (12.8). Under these circumstances, the gain of an individual antenna, assuming both antennas are identical, is given by

$$G = \frac{4\pi R}{\lambda} \sqrt{P_R/P_T} \tag{12.13}$$

where P_R is the power received, and P_T is the power transmitted; R is, of course, the distance between the two antennas.

The quantities in Eq. (12.13) can be measured easily and accurately. Although there may be some uncertainty in measuring the separation of the antennas as to what portions of the antenna should be used, if R is large, the aperture-to-aperture distance indicated in Fig. 12.7 is sufficiently accurate. The quantities P_R and P_T are not measured directly, but instead their ratio is determined. First, the detection system is connected to the receiving antenna, and a reference level is noted on the output meter. Then, the detection system is moved to the transmitter and connected at the output of the calibrated attenuator. The attenuator is adjusted until the output meter reads the same reference level. The ratio of P_T to P_R in decibels is the sum of the coupling of the directional coupler and the attenuation of the calibrated attenuator. This can be expressed as a number and its reciprocal can be used in Eq. (12.13).

As an example of the use of Eq. (12.13), assume the horns which are to be calibrated have an aperture of seven by four inches, and are measured at 9600 megacycles ($\lambda = 1.229$ inches). From Eq. (12.8), R must exceed

$$\frac{2 \times 49}{1.229} = 80 \text{ in.}$$

For simplicity let R be 100 inches. Now with a 20-decibel coupler at the transmitter it is found that two decibels of additional attenuation is needed to bring the power down to the same reference level that was measured at the receiver. Thus $P_T/P_R = 22$ db which, as a number, is 158.5. Its reciprocal is 0.00631. Thus, from Eq. (12.13), the gain of each horn is

$$G = \frac{4\pi 100}{1.229} \sqrt{0.00631} = 80.8 \qquad (12.14)$$
$$\approx 19.1 \text{ db}$$

12.17. PATTERN MEASUREMENTS

It is evident that a picture of the beam pattern of an antenna, like Fig. 12.2, contains information about the beamwidth, side-lobe level, location of sidelobes, location of nulls between lobes, and level of the back lobe. In Fig. 12.2, the three-decibel beamwidth is 20°. (If the information is desired, the ten-decibel beam width can also be noted as 42°.) One sidelobe with peak at about 54° is 14.6 decibels down, and the other major sidelobe, at about 58° on the other side of the main beam, is slightly more than 15 decibels down. Other sidelobes are more than 18 decibels down, and there is no back lobe. Nulls occur at about 30° on each side of the main lobe.

A pattern like Fig. 12.2 is obtained with the set-up shown in Fig. 12.8. The transmitter must be separated from the receiver by at least $2D^2/\lambda$. Care must be taken that between the two antennas there are no objects which will cause reflections which might be picked up in a sidelobe and mistaken for energy in the main beam.

The receiving antenna is mounted on a pedestal which can rotate

360° in azimuth and can also be tilted in elevation to align the two antennas. The motor which runs the pedestal is connected to the turntable of a recorder which duplicates the angle of rotation. The detector output is connected to the pen on the recorder. Thus, as the antenna turns, a plot of its pattern is made automatically.

After the antennas are lined up in both elevation and azimuth, so that the power received is a maximum, the gain of the recorder is adjusted so that it reads full scale when the antennas are in line. The antenna is then rotated through 360° and the pattern is recorded.

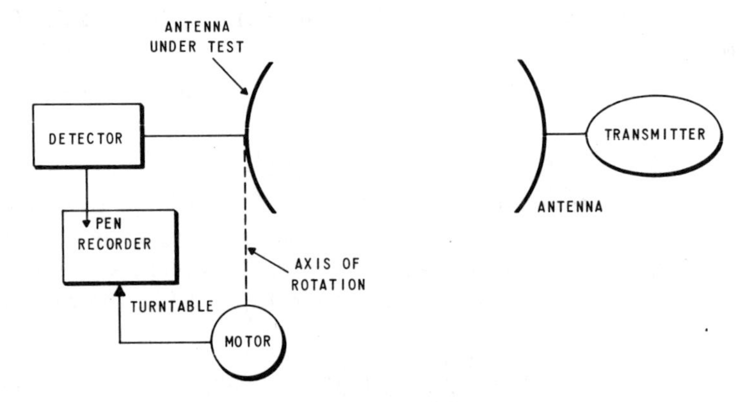

Fig. 12.8. Pattern measuring set-up.

For the E-plane pattern of Fig. 12.2, the antenna must be rotated in the E-plane; that is, both the transmitting antenna and the receiving test antenna must be horizontally polarized. For an H-plane pattern, both antennas are vertically polarized. It is not desirable to try to take either pattern by moving the test antenna in elevation, since at low elevations it might pick up reflections from the ground.

QUESTIONS AND PROBLEMS

12.1. Explain the "Reciprocity Theorem" and its importance in making antenna measurements.

12.2. Define *isotropic antenna, directivity, effective area,* and *antenna gain.*

12.3. Why are the dimensions of antennas often given in wavelengths when the subject of their gain is being discussed?

12.4. Define *vertical polarization, circular polarization, cross-polarization,* and *radiation pattern.*

12.5. A parabolic antenna has tapered illumination (down 10 db at the edges). The diameter is 50 cm and the frequency 10 Gc.
a. What is the beamwidth (assume equal E and H patterns)? *ans.* 5.2°
b. What is the beamwidth if uniform antenna illumination is assumed?

12.6. Explain a method of accomplishing conical scanning and a use that might be made of this operation.

12.7. Distinguish between the Fresnel and Fraunhofer regions and point out why they are of importance when one is making antenna pattern measurements. What is the minimum distance that should be used in checking an antenna similar to the one in problem 5?

12.8. You wish to check the gain of an antenna by using a calibrated standard horn with a gain of ten. The power of the unknown peaks is −20 db. The standard is substituted and the level is −25 db. What is the power gain of the antenna being tested? *ans.* 15 db
How can one avoid square-law detector errors in making gain measurements?

12.9. You have two identical parabolic antennas and wish to check their gain. When they are separated by 50 meters, the power transmitted by one is found to be 100 Mw and the power received by the second .01 Mw. The frequency is 15 Gc. What is the power gain of the antennas? Is this separation sufficient to make the results valid? Explain.

12.10. You are using two antennas in a microwave link; one of these has a gain of 15 db; the second has a gain of ten db. The transmitted power is 100 w. What transmitter power would be required if the antennas were replaced with two antennas having a gain of 40 db each?

12.11. What is the beamwidth of the antenna used for the conical scanning in Fig. 12.4? Make an estimate of the gain of this antenna if the E and H patterns are assumed to be the same.

12.12. The antenna used to obtain the pattern of Fig. 12.2 is transmitting 500 w. A receiving antenna at 0° picks up 10 mw. What would the same antenna pick up if the distance remains the same but it is placed at 55°? At 90°?

12.13. A paraboloid of revolution is to have a power gain of 4000 at a frequency of 10 Gc. What is the minimum diameter required?

12.14. What is the effective area of an isotropic radiator at 5 Gc? *ans.* 2.8 cm^2

12.15. What is the maximum power that can be received over a distance of 1.5 km in free space, with a 1.5 Gc circuit consisting of a transmitting antenna with a gain of 25 db and a receiving antenna with a gain of 30 db? The transmitted power is 200 w.

13

MICROWAVE TUBES

Ordinary triodes, such as the kinds used in broadcast receivers or TV sets, have an upper frequency limit which precludes their use at microwaves. The lead lengths to the electrodes have inductance, which in combination with the interelectrode capacitance determines an upper frequency limit. Actually, even if inductance and capacitance are minimized, the tube will cease to operate at a lower frequency than that determined by these parameters. This is the result of *transit time*, which becomes more important at higher frequencies. Transit time is the time required for electrons to travel between electrodes. If this time is an appreciable part of a cycle, the tube will not operate properly. At low frequencies, an electron leaves the cathode and travels to the anode during the positive part of the grid voltage. At higher frequencies, the electron would leave the cathode when the grid is positive, but before the electron passes the grid, the grid voltage might be negative, or might even go through several cycles.

13.1. DISK-SEAL TRIODES

A special triode for use at microwaves is shown in cross section in Fig. 13.1. The problem of lead inductance and interelectrode capacitance has been solved by making the electrodes metal disks which are sealed to a glass envelope. Thus, the tube is called a *disk-seal* tube. Also, because of its shape, it is sometimes called a *lighthouse* tube.

In order to minimize transit time effects, the interelectrode spacing is reduced inside the tube. Typically, the spacing between the cathode and the grid is only 0.004 inch and between grid and anode is 0.016 inch.

The disks in the disk-seal tube are not leads to the electrodes. Instead, the disks and the electrode structures are parts of reentrant resonant cavities surrounding the interelectrode space. As is illustrated in the cross section of Fig. 13.2, the grid is part of the common wall between the two cavities. The glass envelope of the tube acts simply as dielectric in each cavity and has little effect on operation.

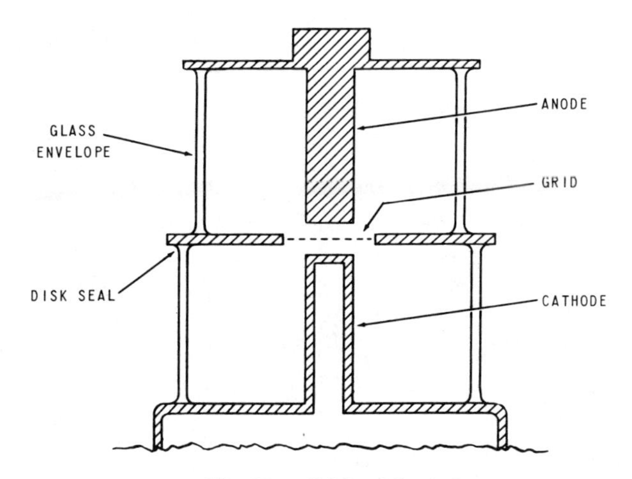

Fig. 13.1. Disk seal triode.

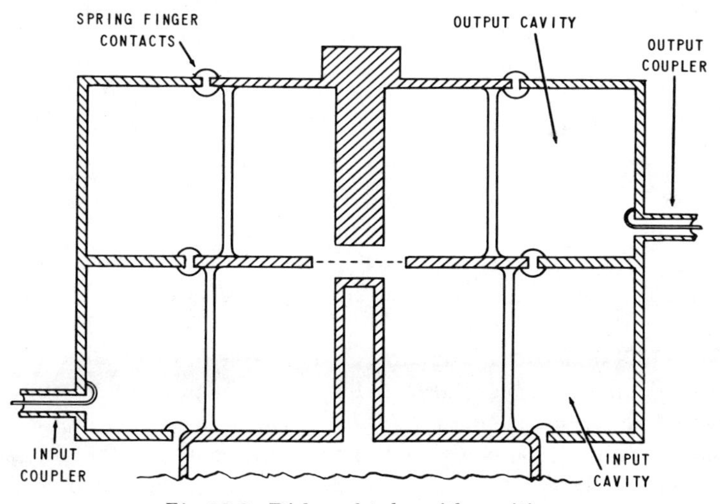

Fig. 13.2. Disk seal tube with cavities.

As an amplifier, the tube operation is similar to that of a conventional triode. A small signal is fed into the grid-cathode cavity by means of the input coupling loop. This signal produces a voltage across the grid-cathode space in the cavity, which in turn controls the number of electrons traveling from the cathode to the plate. Because the grid-anode space is much larger than the grid-cathode space, the resultant change in voltage between grid and anode will be larger than the change between grid and cathode. Thus, the output signal will be amplified. The frequency of the input signal, of course, must be the resonant frequency of the cavities.

The lighthouse tube may be used as an oscillator by supplying feedback from the output cavity to the input cavity. An additional coupling

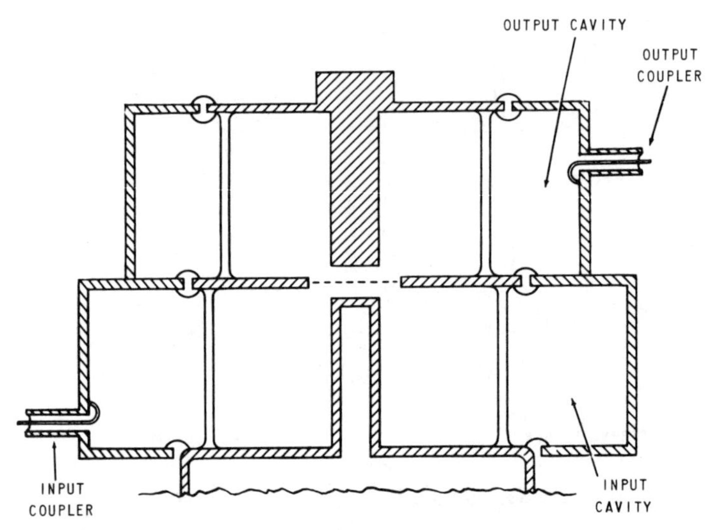

Fig. 13.3. Disk seal tube as frequency multiplier.

loop in the grid-plate cavity removes a small amount of energy and returns it to the input loop. The frequency of the signal at the output is controlled by the resonant frequency of the cavities. By clever mechanical arrangements, the cavities can be tuned to permit variable frequency operation.

Figure 13.3 illustrates the use of a disk-seal tube as a frequency multiplier. It is only necessary to have the output cavity tuned to a harmonic of the input frequency.

Disk-seal tubes have been used at frequencies up to 3000 megacycles. The interelectrode spacings had to be minimized to achieve this. To reduce the spacing further would increase the likelihood of voltage breakdown across the gap.

13.2. INTERACTION

Since the transit time presents a difficult problem in using conventional types of tubes for high frequencies, the microwave tube designers have invented ingenious circuits and structures which utilize the finite transit time as an asset. Basically, the microwave tubes are characterized by some sort of interaction between the electron beam and the r-f signal. There are many structures and methods of doing this but they fall into two broad categories, *local interaction* and *extended interaction.*

Interaction that takes place in a very short time (usually less than a cycle), is termed *local.* Since whatever effect the electrons have on the r-f signal, or vice versa, must happen quickly, the interaction must be intense. Usually this implies a high r-f voltage and a circuit with high loaded Q. Consequently, such tubes are almost always narrow-band in operation. Klystrons are typical of local interaction tubes.

In some tubes, the r-f signal and the electron beam travel together for many cycles. This is *extended* interaction. Since the electron beam and r-f signal are able to interact for a relatively long time and over a long distance, the r-f voltage can be lower, and high Q circuits are unnecessary. This permits broad-band operation. The traveling wave tube is typical of extended interaction tubes.

Occasionally some tubes are described as having intermediate interaction. This implies that there is no firm demarcation between local and extended interaction.

13.3. FEEDBACK

When a tube is to be used as an oscillator, it must have some mechanism for taking a part of the output signal and feeding it back into the input to maintain oscillations. This is called feedback, and in microwave tubes it can occur in at least four ways.

The simplest form of feedback is *external;* it is the same as the feedback in conventional tube oscillators at lower frequencies. An external circuit provides a path from the output of the tube back to the input. A small portion of the output signal is fed back on this external circuit. It is amplified in the tube, and the process is repeated.

Feedback may be supplied by means of a *reentrant circuit.* This is a slow wave structure that feeds back into itself. The r-f signal circulates, and every time it passes the output, a small portion is extracted. During each revolution there must be sufficient amplification to compensate for the amount extracted at the output.

Beam reentrant feedback occurs when the electron beam circulates through the interaction region. Frequently both types of reentrancy are used in the same tube.

In some tubes, modulation on the electron beam gives rise to an r-f signal which travels backwards with respect to the beam. This is called a backward wave mode of operation. When the r-f signal arrives at the electron gun, it remodulates the beam, and the process is repeated. This is called *interactive feedback*.

13.4. VELOCITY MODULATION

In ordinary vacuum tubes, the beam of electrons from the cathode to the anode travels at a constant velocity, but the amplitude or number of electrons in the beam is varied and controlled by the grid voltage to produce amplitude modulation. In some microwave tubes, the velocity of the electrons is varied and controlled, but the total number of electrons in

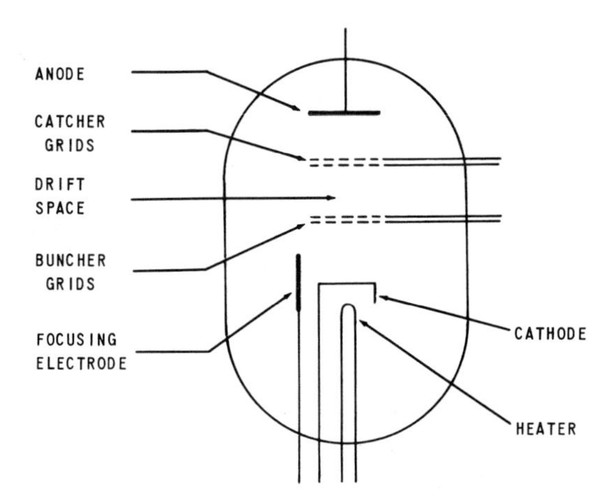

Fig. 13.4. Velocity modulation.

the beam remains constant. This is called *velocity modulation*. It is useful because it utilizes transit time to produce amplification and then produces amplitude modulation in the output.

A method of producing and using velocity modulation is shown in Fig. 13.4. The cathode produces a beam of electrons which is then focused by a focusing electrode. The assembly of heater, cathode, and focusing electrode is called an *electron gun*, and similar assemblies are used in cathode ray tubes as well as in many microwave tubes. The electrons in the beam travel towards the anode which has a voltage on it which is positive with respect to the cathode. An r-f signal is connected across a pair of *buncher grids* which are closely spaced so that transit time is a minimum. The r-f voltage on these grids is isolated from the cathode and anode so the electrons travel through without difficulty. When an electron is between the two buncher grids, however, its velocity will be affected by the relative voltages on the two grids. In other words, if the second grid is positive with respect to the first, the electron will be accelerated; however, if the r-f cycle is at a point where the second grid is negative with respect to the first, the electron will be slowed. After the electrons have been speeded up or slowed down they travel at their new velocities in the drift space. Here there are no additional voltages acting on the electrons, so the fast electrons overtake the slow ones. Thus, groups or *bunches* of electrons are formed. At some point in the drift space, this bunching reaches a maximum and is said to be *complete.* If the space were longer, the fast electrons would begin to pull away from the slow ones, while if it were too short, they would not have caught up completely.

A second pair of closely spaced grids is placed at the point of complete bunching. As the bunches of electrons pass between the grids, a voltage appears across them proportional to the number of electrons in the space. Since the number of electrons is varying at the r-f rate, the voltage across these *catcher grids* also varies at the r-f rate. Because the bunching at the location of the catcher grids is much tighter than the bunching at the start (at the buncher grids), the r-f voltage at the catcher grids is much greater than the original r-f voltage across the bunching grids. The amplification is of the order of 1000.

The interaction at the catcher grids is local since it takes place over a very small portion of the electron beam. Actually, although there is a long drift space, the interaction between the electron beam and the r-f circuitry occurs just twice—both times in the narrow confines between the pairs of grids.

13.5. MULTICAVITY KLYSTRON

The arrangement of elements illustrated in Fig. 13.4 might be suitable at low frequencies, but at microwaves it would be difficult to apply the r-f signal to the buncher grids without some additional circuitry. In order to avoid the reactances associated with wire leads, the grids in the circuit are made parts of resonant cavities. The complete tube is called a *multicavity klystron* and is illustrated schematically in Fig. 13.5.

The klystron utilizes velocity modulation. Electrons leave the gun and enter the *buncher cavity* where an r-f signal produces velocity modulation. When the electrons enter the *catcher cavity*, the velocity modulation produces an amplified r-f signal there (as was explained in the preceding section). If the tube is to be used as an amplifier, the input signal is loop-coupled to the buncher cavity, and the output is loop-coupled out of the catcher cavity.

The klystron can also be used as an oscillator by supplying external feedback. A second output loop is supplied in the catcher cavity, and the sampled output is fed back into the input line.

The klystron is a narrow-band tube, but it may be tuned in frequency by varying the resonant frequency of the cavities. This is usually done mechanically by making one wall of each cavity a flexible diaphragm.

The klystron can also be used as a frequency multiplier. A velocity-modulated beam is rich in harmonics, so multiplication factors of ten or

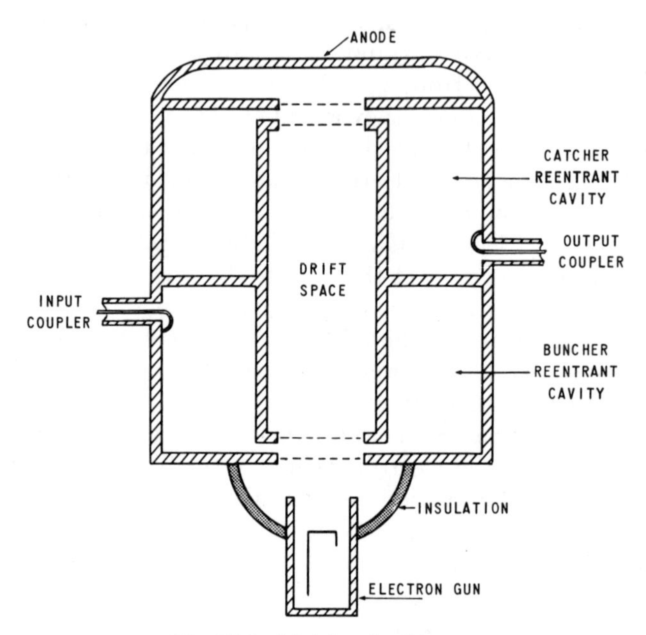

Fig. 13.5. Multicavity klystron.

more are readily obtainable. It is only necessary to have the output catcher cavity tuned to the desired harmonic.

13.6. REFLEX KLYSTRON

A *reflex klystron* uses only one cavity as both the buncher and catcher. The electron beam first goes through the cavity and is velocity-modulated. Then it is reflected and passes back through the cavity. The arrangement is illustrated in Fig. 13.6. The cavity is also the anode.

Since the outer conductor of the output line is tied directly to the cavity, the cavity is usually at ground potential. The cathode is then made negative with respect to ground. As in the multicavity klystron, the electron beam from the gun goes through the cavity and is bunched. Instead of a second cavity, the drift space is followed by a *reflector* or *repeller*, which is simply an electrode which is negative with respect to the cathode. The electrons do not actually reach the repeller, but are reflected sooner by the negative voltage. The electrons then enter the cavity which now acts as a catcher for these electrons, while at the same time it is a buncher for new electrons coming from the gun.

The drift space in the tube is called the *repeller* space and is traversed twice by the bunched electrons. The point at which the electrons turn is controlled by the repeller voltage. The more negative the voltage, the sooner the electrons turn. For maximum power out, there should be

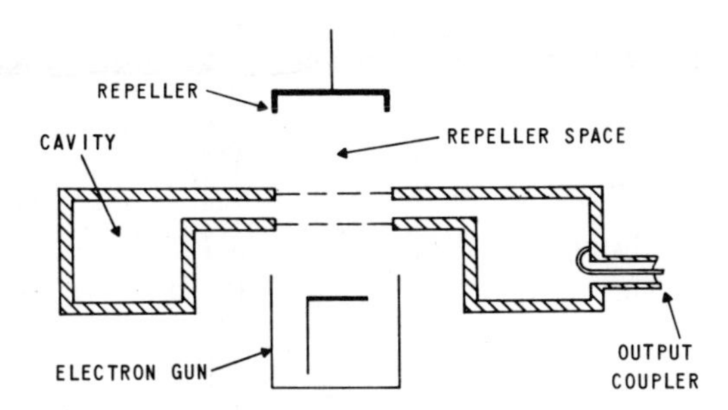

Fig. 13.6. Reflex klystron.

complete bunching of the reflected beam at the catcher; thus, the repeller voltage must be adjusted to accomplish this bunching.

Frequency is controlled primarily by tuning the cavity, but the repeller voltage has a secondary effect. Varying the repeller voltage varies frequency, over a narrow range, as well as power output.

The reflex klystron is used as a local oscillator in microwave receivers and as a signal source for general laboratory use. The interaction is local as in the multicavity klystron, and the feedback is reentrant.

13.7. MAGNETRON

The *magnetron* was the first microwave high power oscillator. It is simply a vacuum tube with two elements, a cathode and an anode, and several resonant cavities. A cut-away view displaying the physical arrangement is shown in Fig. 13.7, and a cross section is illustrated in Fig. 13.8. The cathode is a rod through the center of the cylindrical tube, and the anode is a solid block containing the resonant cavities. The space between the cathode and the anode is the *interaction space.*

Not shown in the figures, but an important part of the tube, is an applied magnetic field with flux lines parallel to the axis of the cathode; that is, the poles of the magnet are at each end of the cathode, but remain outside the vacuum envelope. The magnetic field is usually supplied by a fixed horseshoe magnet, but it may also be supplied by an electromagnet.

When electrons leave the hot cathode, they proceed radially to the anode. However, the applied magnetic field causes the electrons to curve in their paths. With a critical magnetic field, the electrons would describe arcs which just graze the anode and bend back toward the cathode. At higher values, the electrons would spiral in the interaction space and never stray far from the cathode. In practice, the magnetic field is near the critical value.

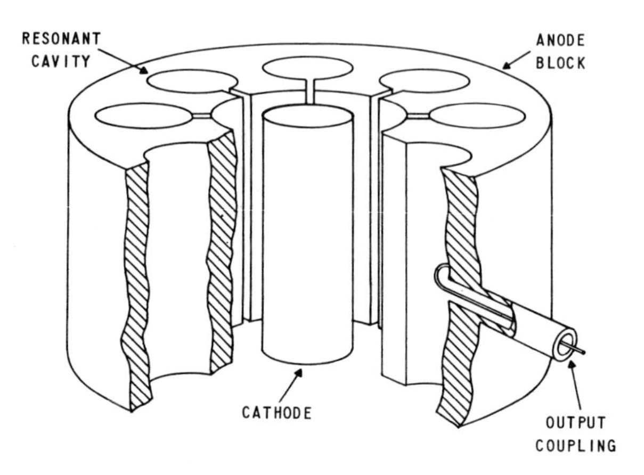

Fig. 13.7. Magnetron—cutaway view.

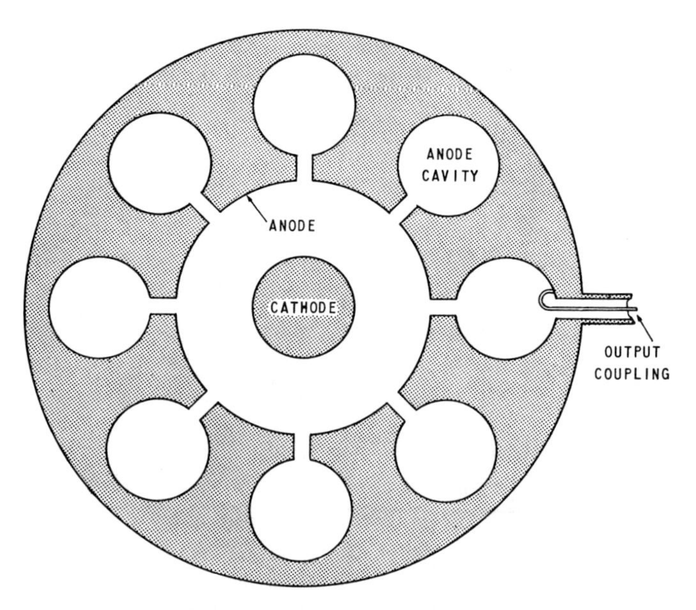

Fig. 13.8. Magnetron cross section.

In order to understand the operation of the magnetron, it must be assumed that oscillations are started so it can be shown that the oscillations will be maintained. (This is the usual form of oscillator explanation.) If there are r-f oscillations in the resonant cavities, the electrons in the interaction space will interact with the r-f signal every time they pass part of the anode between cavities. As the electrons move out from the cathode, they gather energy. When an electron approaches a part of the anode just as that part is entering the positive part of the r-f cycle, the

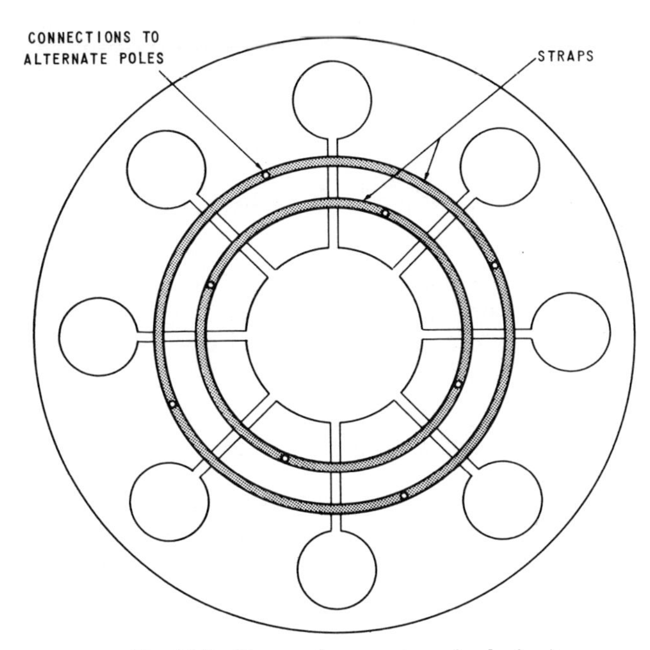

Fig. 13.9. Strapped magnetron (end view).

electron's approach will give an additional push to the positive-going signal. In doing so, energy is transferred from the electron to the r-f signal, and the electron returns to a smaller orbit. This action is repeated as the d-c anode voltage again pulls the electron away from the cathode.

Every time an individual electron approaches the anode "in phase" with the r-f signal, it completes a cycle. The total phase shift around the periphery of the anode must be some multiple of 2π. The anode may have two cavities, or eight as shown in Fig. 13.8, or, in fact, any convenient even number. The more cavities there are, the more frequencies exist at which the magnetron will operate. Thus, an electron can give energy to one part of the anode and drop to a smaller orbit, and, on its return to the anode, it may approach the next adjacent section or a section farther away. These are called different modes of operation and result in different frequencies. In the arrangement shown in Fig. 13.8, these different frequencies are close together, and as a result, the magnetron oscillates sometimes at one frequency and sometimes at another. For the dominant mode, adjacent anode poles have a phase difference of π radians. This is called the *π mode.*

If the frequencies of the different modes of operation are far apart, the magnetron will not have a tendency to change from one to another so readily. One method of achieving frequency separation is by *strapping,* illustrated in Fig. 13.9. Two rings are connected to the ends of the anode, with each ring connected only to alternate anode poles. At the π mode, each ring is at a uniform potential, but the two rings have opposite potential and thus present a capacitive loading to the cavities, which lowers the frequency of this mode. For other modes, each ring is not at a uniform potential, so that current flows in the rings. This places an inductive loading on the cavities which raises the frequencies for other modes. The wanted π mode is thus separated from all the others.

The interaction in the magnetron is local, but occurs frequently. It is usual then to call it intermediate. The feedback is both interactive and reentrant.

13.8. FREQUENCY CONSIDERATIONS

All microwave oscillators are affected by the voltages applied to the electrodes and by variations in the load impedance. These effects are called *pushing and pulling.* When the frequency of an oscillator changes because the anode voltage (of a magnetron, for example) or the repeller voltage (of a klystron) drifts or is changed purposely, the tube is said to be *pushed.* (It is also correct to say the frequency of the tube is pushed.) The frequency will also change if the load that the tube feeds into varies. In this case, the tube (or the frequency) is said to be *pulled.* Variations in either phase or amplitude can affect the frequency. To a large extent the resonant cavity or other frequency-determining element inside the tube sets the frequency of oscillation. However, just as in a multiple-tuned circuit, the voltages applied and the load into which the tube is fed can affect this frequency.

The pushing effect can be used to advantage. A klystron's frequency will not remain stable with time even if the voltages on the electrodes are held constant. The cumulative effect of internal heating causes parts to change dimensions which causes a subsequent change in frequency. If a sample of the signal is fed through a fixed reference cavity external to the klystron, the phase shift through the cavity will vary directly with the deviation in frequency from cavity resonance. The phase shift, then, is a measure of the frequency error. By suitable circuitry this is converted to a d-c voltage correction to be applied to the repeller of the klystron to bring the tube back to the proper frequency. Thus, pushing can be used to maintain stable frequency.

In some systems, it is desirable to frequency-modulate a magnetron. Typically, the frequency is saw-tooth modulated. When an echo is

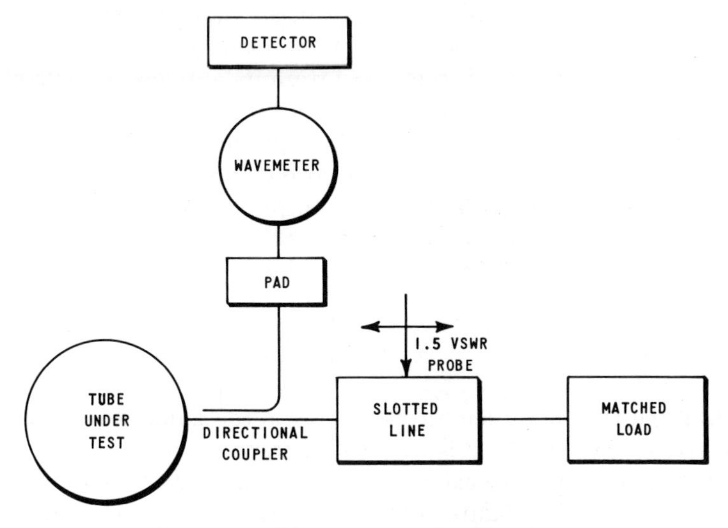

Fig. 13.10. Measurement of pulling figure.

received from a target, it will be at a frequency which left the magnetron some time earlier. The difference in frequency between the received echo and the present magnetron frequency is a measure of the time that it took for the signal to go out and return. The time can be converted to the distance of the target.

By proper design of the anode, it is possible to build a magnetron whose frequency is linearly dependent on the anode voltage. These voltage-tunable magnetrons (or VTM) can be used in systems requiring rapid change of frequency, remotely controlled frequency output, or frequency modulation. Voltage-tunable magnetrons have been built to swing over an octave or more.

The *pulling* figure of a tube is a quantitative measure of the effects of load variations. The pulling figure is measured in megacycles and is the total frequency shift in the oscillations when a load with a VSWR of 1.5 is varied through all phases. This may be measured as shown in Fig. 13.10. The output of the tube is fed into a slotted line which is terminated in a matched load. A piece of dielectric is inserted into the slot to produce a VSWR of 1.5. Then this piece is slid along the slot so that its position is changed by at least 360°. The frequency is monitored continuously, and the *total* frequency change is the pulling figure.

Klystrons are less susceptible to pulling (that is, they have lower pulling figures), than magnetrons. However, when using a klystron as a signal source in the laboratory, it is necessary to pad it, preferably with an isolator, in order to eliminate errors which might arise from pulling.

When a magnetron is far from its load, a small change in frequency produces a large change in the phase of the load. This is called the *long line effect*. If the magnetron is separated from its antenna by several feet, as in a typical system, and if the magnetron is frequency modulated, this long line effect can produce frequency jumping in the magnetron in place of the continuous frequency variation required. The farther the antenna is from the magnetron, the more pronounced is the frequency jumping. For a given pulling figure and VSWR, there is maximum safe length at which there will be no pulling. This is given by

$$L = \frac{130}{(\lambda_g/\lambda)P(r^2 - 1)} \tag{13.1}$$

where L is the maximum length in feet, λ_g is the guide wavelength, λ is the free-space wavelength, P is the pulling figure in megacycles, and r is the VSWR of the load. For example, in a system at 9350 megacycles, the magnetron has a pulling figure of 20 megacycles, and the antenna VSWR is 1.6. Then from Eq. (13.1),

$$L = \frac{130}{1.40 \times 20 \times 1.56} \approx 3 \text{ ft} \tag{13.2}$$

Thus, if the magnetron can be placed three feet or less from the antenna, there will be no frequency skipping. If the antenna VSWR can be reduced, the separation would be greater. Since in practice three feet is not enough, it is customary to place a ferrite isolator right after the magnetron so that the VSWR it sees is reduced to an acceptable figure.

Equation (13.1) may be rewritten as

$$r^2 = \frac{130}{(\lambda_g/\lambda)PL} + 1 \tag{13.3}$$

Here L is the actual distance between magnetron and load, and r is the maximum VSWR that can be tolerated. For example, assume an air-filled coaxial line 20 feet long separates the magnetron and antenna, and assume also that the pulling figure is fifteen megacycles. Since $\lambda_g = \lambda$ in an air-filled coaxial line, then

$$r^2 = 1 + \tfrac{130}{300} = 1.433$$

or (13.4)

$$r \approx 1.2$$

If the antenna VSWR is 1.6, it must be reduced to 1.2 by an isolator. Eight decibels of padding is required to reduce a 1.6 VSWR to 1.2. Thus the isolator should have at least eight decibels of isolation and, of course, should have low loss in the forward direction.

13.9. TRAVELING-WAVE TUBE

In a klystron which is used as an amplifier, the interaction is concentrated in a small space, and the result is a narrow-band amplifier. If the interaction could somehow be extended over a long path, it would not be necessary to have it so intense, the circuit could have a lower Q and be broadband. In order to have extended interaction, the electron beam and the microwave signal must travel together at the same rate of speed. Unfortunately, the r-f signal normally travels at the speed of light while the electrons are much slower. However, the r-f signal can in effect be slowed down by making it spiral around the electron beam. The amount of slowing down is determined by the pitch of the spiral.

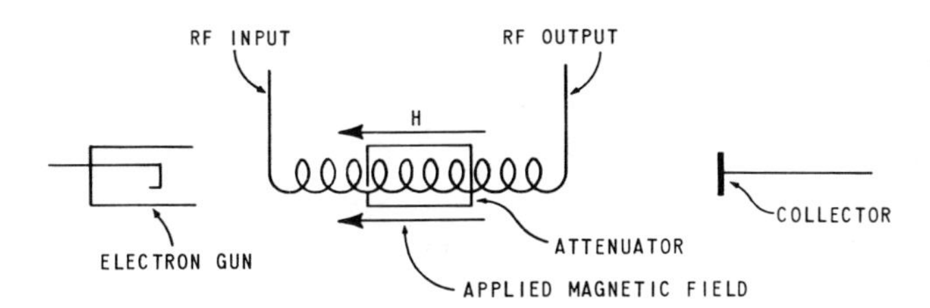

Fig. 13.11. Traveling-wave tube.

Figure 13.11 illustrates the elements of a *traveling-wave tube*, or TWT. An electron gun emits a beam of electrons which, passing through the center of a wire helix or spiral, travel to the collector at the far end. The microwave signal is introduced on the helix at the gun end and travels along the helix to the collector end. Although it travels much faster than the electrons in the beam, the axial motion of the r-f signal is the same as that of the average speed of the electrons.

There are many r-f cycles in the length of the tube. At the points in the tube where the r-f voltage is positive, nearby electrons are accelerated. Where the r-f voltage is negative, the electrons are slowed down. Thus, bunches of electrons are produced all along the tube. As these bunches of electrons move toward the collector, their charges induce a voltage in the helix which adds to the electromagnetic wave there; that is, the electromagnetic wave is amplified. Since the interaction takes place along the whole length of the tube, there is continuous amplification. The gain of the TWT is proportional to its length.

The input and output couplers are difficult to match perfectly over a wide frequency range. Consequently, it is possible for a reflection off the output coupler to return through the tube to the input. Here part of it is reflected again and is amplified as it travels toward the output. This results in an unwanted oscillation, which can be prevented by inserting an attenuating material such as graphite in the tube. The attenuator, of course, reduces the gain of the tube, but it does prevent oscillations.

One of the problems in a traveling-wave tube is the difficulty of keeping the electrons from scattering. The electrons all have the same charge and tend to repel each other. Ordinarily, the beam would break up long before the electrons reach the collector. This is corrected by applying a strong axial magnetic field to focus the electrons, as is indicated in Fig. 13.11. This magnetic field can be supplied by an electromagnet or by a series of disc-like permanent magnets surrounding the tube.

Traveling-wave tube amplifiers cover an octave bandwidth with gains of 40 decibels or more. They have low noise figures, some as low as six decibels, so that they can be used in front of a microwave receiver to improve sensitivity. Klystron amplifiers, on the other hand, are narrowband and have noise figures of about 25 decibels. The klystron amplifier will handle much higher powers and can be used in a transmitter chain, where broad bandwidth and low noise figure are not important.

13.10. BACKWARD WAVE OSCILLATOR

A *backward wave oscillator*, or BWO, resembles a TWT because it has an electron gun which fires a beam of electrons through a long helix to a collector. A strong axially-magnetic field is also necessary to focus the electron beam. The BWO is larger in diameter and somewhat shorter in length than the TWT. The microwave signal travels backwards in a backward wave oscillator, and the output is thus near the gun end of the tube.

The electron beam in a BWO is a hollow pencil of electrons very close to the helix around it. The helix is made of flat-wire tape. There is a capacitive interaction between the electrons and the helix which produces the usual bunching. In this tube, the backward traveling wave produces bunches of electrons which in turn reenforce the wave. The interactive feedback is continuous and the interaction extends over the whole length of the tube.

The frequency of oscillation depends on the velocity of the electrons, since at different speeds, they reenforce different frequencies. The velocity of electrons is controlled by the cathode-to-helix voltage. Backward wave oscillators are thus voltage-tunable. Because of the extended interaction, frequency ranges greater than five to one are achievable.

13.11. SPURIOUS SIGNALS

Ideally, a microwave oscillator should emit only one frequency. Unfortunately this is never the case, and all oscillators produce some unwanted signals which are called *spurious signals.* The *main signal* put

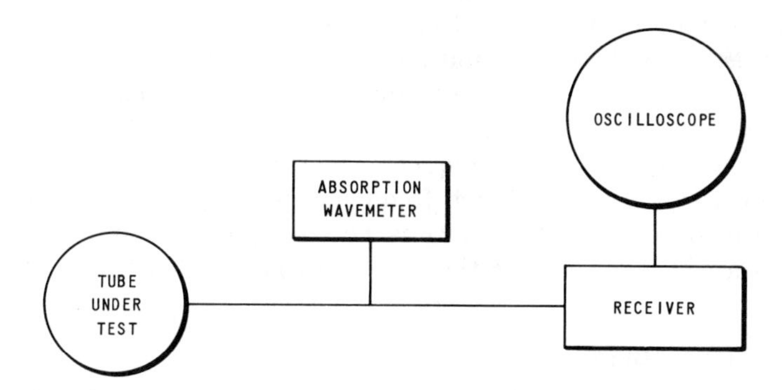

Fig. 13.12. Measurement of spurious signals.

out by a tube is the desired signal and is the signal with highest output for a given set of operating conditions. The level of spurious signals are specified in decibels below the main signal. For example, if a magnetron puts out a 3000 megacycle signal of one watt and a spurious signal at 4000 megacycle of 0.001 milliwatt, the spurious signal is 60 decibels down. If the spurious signals are below the sensitivity of the detecting apparatus, it is proper to say that there are no spurious signals detectable and to state the sensitivity of the detector.

Spurious signals may be detected and measured by the set-up of Fig. 13.12. The output of the tube is fed through an absorption wavemeter to a microwave receiver. First the receiver is tuned to the main frequency of the tube, and the wavemeter is tuned through this frequency. A dip in receiver output will be noticed as the wavemeter is set at resonance for the main frequency. The wavemeter is now detuned, and the receiver is slowly tuned over the frequency band of interest. Several responses are usually observed. As the receiver is tuned, the indications on the oscilloscope move across the face, some from right to left, and others from left to right. If the response moves in the same direction as the main signal moved, when the dial is turned in the same direction, the signal may be spurious. If the response moves opposite to the motion of the main signal, it can be ignored since it is an extraneous signal formed in the receiving system.

If the signal does move in the correct direction, it is not necessarily spurious. The receiver is set to receive such a signal, and the absorption wavemeter is tuned to the *main signal.* If the response is affected by this tuning, the signal is not spurious, but extraneous; that is, it is caused by the main signal mixing with some frequency in the detection system. If tuning to main frequency resonance has no effect on the received response, the wavemeter should be tuned to the frequency of this response. If the response is affected, it is a spurious signal coming from the tube. The amplitude of the spurious signals and of the main signal can be determined by replacing the tube being examined with a calibrated signal generator and feeding each frequency in turn into the receiver. If the output of the generator at each frequency is adjusted to give the same response on the oscilloscope as was noted with the tube in place, the calibration will then indicate the power output at each frequency.

13.12. OTHER TUBES

There are many microwave tubes that are not listed above, and the list is continually growing. The carcinotron, amplitron, phantastron, and many others, usually with the suffix "-tron," all have special capabilities or advantages. However, their principles of operation are all similar to those described in this chapter. There is some sort of interaction and, in oscillators, some sort of feedback. If the principles of interaction and feedback are understood, it is not difficult to understand the operation of any new tube that is developed.

PROBLEMS AND QUESTIONS

13.1. Briefly describe the construction and operation of the lighthouse tube.

13.2. Distinguish between extended and local interaction. How are they related to bandwidth?

13.3. Explain four methods of obtaining feedback to convert an amplifier to an oscillator.

13.4. Describe how velocity modulation of a beam is obtained in a klystron amplifier. How does a reflex klystron differ from an amplifier klystron?

13.5. What is the significance of the term "critical magnetic field" as it is used in connection with magnetrons? What is the purpose of strapping in a magnetron?

13.6. Define the terms "pushing" and "pulling" as they are applied to microwave oscillators. How can the pulling effect be used to advantage?

13.7. What is "the pulling figure?" Explain its relation to VSWR and maximum line length.

13.8. The frequency of a magnetron is 5.4 Gc. The pulling figure is 10 Mc. The VSWR on the line is 1.2. What is the maximum distance the load may be placed from the oscillator? Suppose that the distance found in this situation is not great enough for the given situation; if it is not practical to reduce the VSWR what can be done to avoid frequency jumping?

13.9. What is the most important advantage of the TWT over the klystron amplifier? Why is the klystron used in many applications?

13.10. What is a BWO? How is the frequency controlled?

13.11. Draw a block diagram and explain how you would measure the spurious signals being emitted by a transmitter?

13.12. Explain how a long transit time can make the input impedance of a conventional tube have a resistive as well as a reactive component even though the grid signal is less than the negative bias.

13.13. Draw a block diagram using a reflex klystron as the local oscillator in a microwave receiver. Show how the frequency could be controlled to keep the receiver tuned properly.

appendix

MICROWAVE SCHEMATICS

Microwave schematic diagrams combine block diagrams and pictorial symbols. Whenever there is apt to be ambiguity or doubt, the name of the component or element is written in a block instead of using a symbol. A section of a transmission line is simply a straight line connecting other elements. The accompanying table lists the symbols used to specify the type of line and to indicate other components.

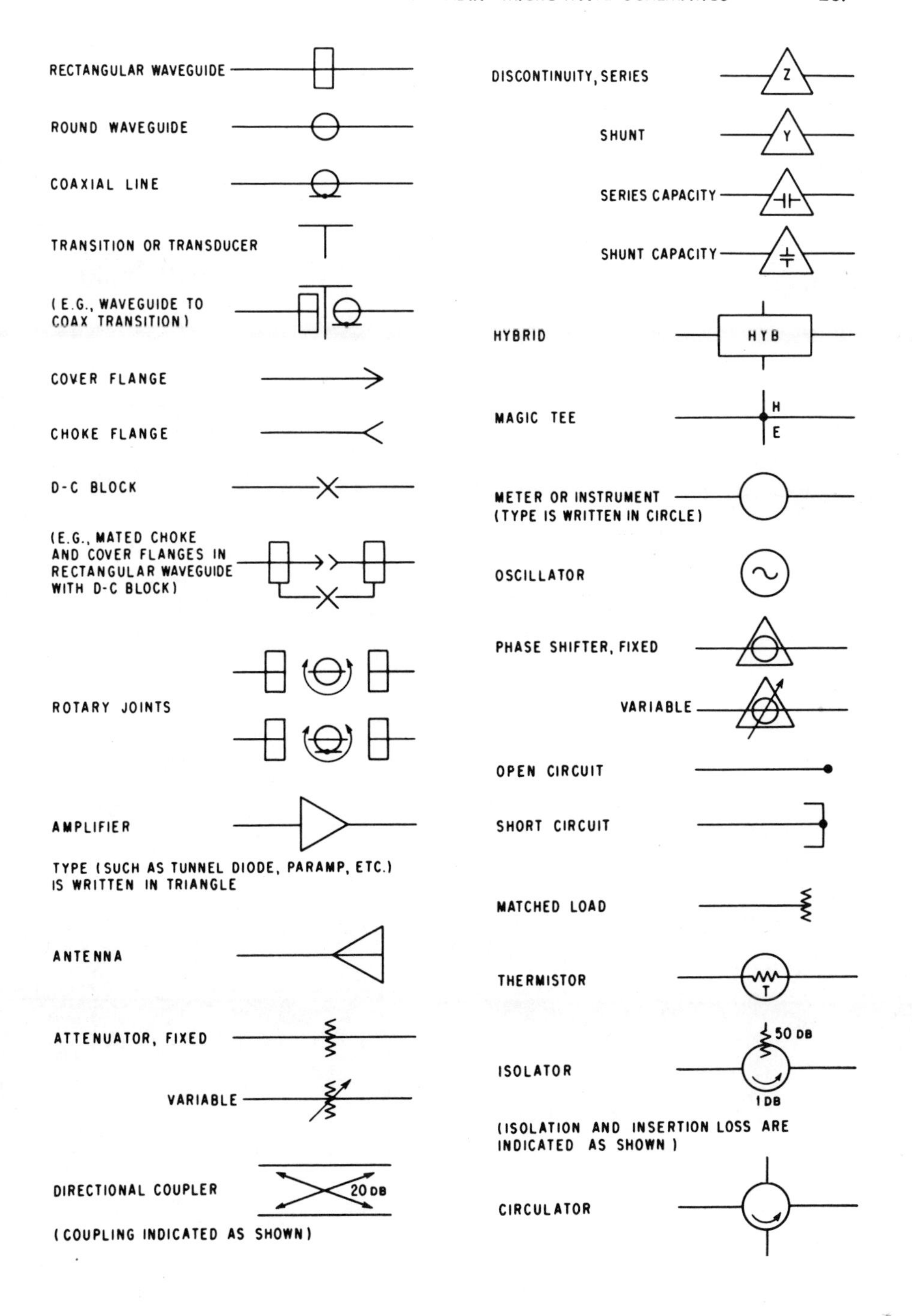

RECTANGULAR WAVEGUIDE
ROUND WAVEGUIDE
COAXIAL LINE
TRANSITION OR TRANSDUCER
(E.G., WAVEGUIDE TO COAX TRANSITION)
COVER FLANGE
CHOKE FLANGE
D-C BLOCK
(E.G., MATED CHOKE AND COVER FLANGES IN RECTANGULAR WAVEGUIDE WITH D-C BLOCK)
ROTARY JOINTS
AMPLIFIER
TYPE (SUCH AS TUNNEL DIODE, PARAMP, ETC.) IS WRITTEN IN TRIANGLE
ANTENNA
ATTENUATOR, FIXED
VARIABLE
DIRECTIONAL COUPLER
20 DB
(COUPLING INDICATED AS SHOWN)
DISCONTINUITY, SERIES
Z
SHUNT
Y
SERIES CAPACITY
SHUNT CAPACITY
HYBRID
HYB
MAGIC TEE
H
E
METER OR INSTRUMENT
(TYPE IS WRITTEN IN CIRCLE)
OSCILLATOR
PHASE SHIFTER, FIXED
VARIABLE
OPEN CIRCUIT
SHORT CIRCUIT
MATCHED LOAD
THERMISTOR
T
50 DB
ISOLATOR
1 DB
(ISOLATION AND INSERTION LOSS ARE INDICATED AS SHOWN)
CIRCULATOR

INDEX